The Limits of Science

WPI Studies

Lance Schachterle
General Editor

Vol. 19

PETER LANG
New York • Washington, D.C./Baltimore • Boston • Bern
Frankfurt am Main • Berlin • Brussels • Vienna • Canterbury

Ernest Krausz

The Limits of Science

PETER LANG
New York • Washington, D.C./Baltimore • Boston • Bern
Frankfurt am Main • Berlin • Brussels • Vienna • Canterbury

Library of Congress Cataloging-in-Publication Data

Krausz, Ernest.
The limits of science / Ernest Krausz.
p. cm. — (WPI studies; vol. 19)
Includes bibliographical references and index.
1. Science—Philosophy. I. Title. II. Series.
Q175.K8726 501—dc21 99-30286
ISBN 0-8204-4537-1
ISSN 8097-926X

Die Deutsche Bibliothek-CIP-Einheitsaufnahme

Krausz, Ernest:
The limits of science / Ernest Krausz.
–New York; Washington, D.C./Baltimore; Boston; Bern;
Frankfurt am Main; Berlin; Brussels; Vienna; Canterbury: Lang.
(WPI studies; Vol. 19)
ISBN 0-8204-4537-1

CONTENTS

PREFACE

I have been contemplating and tentatively planning to write a book about science for many years. The idea has germinated over a very long period. Having been trained as a social scientist, and having had a keen interest in the philosophy of science, I began exploring the natural sciences from an epistemological perspective. This led me to the main topic of interest of this book: a basic understanding of how science goes about explaining reality—the world we live in. Once immersed in this endeavor I began to focus specifically on the question whether science, with its rational methods, will be able to make, what Isaiah Berlin calls, "indefinite further progress in the unravelling of the mysteries of nature" (Berlin, 1981). In other words, is scientific knowledge truly unlimited?

In writing this book the opportunity arose to provide, in the first place, a grasp of scientific concepts, theories and processes even for readers with no scientific background at all. I have attempted to define the scientific terms, theories and philosophical ideas concerning science, which I introduced into this text. This book could, therefore, also serve students who are interested in science, whether with the intention of entering a specific scientific field, or just wishing to widen their knowledge. All in all, if it is not too much to claim, my work not only explains what science is about, but also presents an introductory assessment of the scientific achievements of the modern age, especially during the twentieth century.

It will be up to the reader to judge whether my presentation is a balanced one. It was certainly my aim to provide a balanced account, which I tried to achieve by showing the failures and uncertainties of science as well as its tremendous achievements. I definitely do not belong to that group of social scientists and relativists whom Roger Newton sees as the antagonists of the natural sciences; those who portray the results of scientific investigations of nature and the external world as "simply narratives, like myths and fairy tales, or the outcome of social agreements" (Newton, 1997). But neither do I believe in the infallibility of science; and I do question the ability of science to attain apodictic knowledge leading to some kind of final truth. After all, science is created by man as he confronts nature, and so it carries the limitations of the human mind. Roger Newton himself points out that: "The principal difficulty arises from the limitations of our language, which is tied to the scale of everyday life and seems ill-adapted to the micro world closed to our senses" (Newton, 1997). The problem of how language and culture affect perception and scientific work is obvious. But even more problematic is the limited capacity of the human mind to conceive and understand the macro world: the universe (or universes), its (their) origin, existence and future.

That there are limits to what science can attain in knowledge is now realized. But I still considered it necessary to convey a strong message in this book, that science is not the 'know-all' that it is often imagined to be. I thought it made good sense to present science as it is, rather than an idealized version of it. Thus, one arrives at the inevitable conclusion that science cannot provide answers even to basic questions such as how mass is generated or how living organisms originated, and it does not provide a proper understanding of human consciousness. Whether such ultimate questions are of more or less importance than the actual problems solved by science, or the explanations it gives to many hitherto unknown or little understood phenomena is quite another matter. The fact is that we constantly arrive at the same limits of ultimately unanswered questions or ultimately hidden causes. Will science ever be able to break these limits? To put it another way, can science overcome the inscrutability of nature? I think not. But scientific work will go on and many new successes will be registered, judging by scientific performance so far. These successes will be deservedly applauded. Yet the end of the story will forever remain unknown—an unsolved mystery.

ACKNOWLEDGMENTS

I wish to express my indebtedness to many people in connection with the writing of this book. I am grateful to my wife Gillian for her constant support and to my children for their professional advice in diverse areas: my daughter Sara (structural biology), my daughter Miriam (economics), my son David (political science and media studies), and my son Benjamin (Talmud and philosophy). I also wish to mention other members of my close family who have been most supportive: my son-in-law Dr. Nathan Cohen and my daughters-in-law Michal and Orah.

My gratitude is due to a number of colleagues at Bar-Ilan University who commented on various parts of the manuscript; but, of course, they are not responsible for any of the shortcomings of this book: Professor Cyril Domb FRS (theoretical physics), Professor Sanford Sampson (cellular physiology), Professor Aryeh Hillman (economics), Professor Harry Friedmann (chemistry), and Professor Daniel Algom (psychology). I also wish to record my gratitude to Professor Lance Schachterle, the Series Editor, and to Owen Lancer, the Acquisitions Editor at Peter Lang, who have made this publication possible. In the final stages of preparing this manuscript I received very helpful comments from Professor John M. Wilkes, of the Social Science and Policy Studies Department at the Worcester Polytechnic Institute, Massachusetts. For this I am most grateful to him.

I carried out much of the work for this book within the research framework of the Institute for Community Studies at Bar-Ilan University, Israel, whose support I greatly appreciated all along. A good deal of the researching and writing was also done on sabbaticals at the School of Social and Human Sciences, City University, London. For enabling me to spend my sabbaticals at City University I wish to thank Dr. Stephen Miller, the Dean of the School and Pro-Vice-Chancellor of the University.

I shall single out a few of the libraries which were most helpful: The British Library of Political and Economic Science at the London School of Economics, the library of the University of

Newcastle-upon-Tyne, England, and the library system at Bar-Ilan University: particularly the Philosophy library and the Physics library.

Finally, my thanks are due to Dr. Esther Guggenheimer-Furman, Menucha Kutner and Sarah Leslau who, at different stages, prepared my manuscript so expertly for publication.

Ernest Krausz
March, 1999

CHAPTER ONE

THE DEVELOPMENT OF MODERN SCIENCE

Some Basics

It would be too simple, and not very enlightening, to state that science or the scientist deals with reality. In our everyday life all of us are engaged constantly in such activity; namely, we confront reality and in our own ways we even attempt to understand it. But the scientist has the definite aim to explain systematically the workings of the world around us, something the layman need not do and presumably could not do since he lacks the training, the tools and the methods which are available to the scientist.

Another distinction worth making is between science and fiction. The latter openly stems from the fancies and flights of the imagination; extrapolations not completely detached from the real world but extending beyond the known. Science also relies heavily on imagination and intuition but is operative in a far more constrained framework. When the world around us is described by creating categories, inventing symbols and formulas, taking measurements, it is not all factual description. There is a creative process of simplification and consolidation going on in an attempt to explain the dynamics of our lives and universe around us by means of theories. These are supposed to explain how things are caused, how they develop, how they change. Science even claims to have uncovered, and that it will continue to uncover, some of the rules which govern the existence of our universe and our lives.

Fundamentally, however, the aim of science is to be based on factual information or to refer back to such information (Kemeny, 1959). The weighing of factual evidence applies to all fields of inquiry: whether it is the physical, biological or social world—the galaxies of an expanding universe, the minute viruses which attack living organisms (including humans), or the social relationships or psychological condition of human beings.

The process of observation is a most complex one. The process is based on a triangular connection between the objects or happenings, called facts, phenomena or events of an external reality, the physical sensory contacts of the observer with that reality, and the workings of his brain/mind in the process of perception. The pen a person wishes to write with is a bit of the external reality; when he sees it and moves it, a physical connection is created between this object, occupying a space in that reality at some point of time, and the observer; and his perception, or awareness, of the pen is due to a mental process taking place in his brain. Observation, in other words, is based on perception. However, there is a fourth dimension in the process of human observation: social interaction and especially communication. The meaning of the observation, not simply the fact that something is perceived, depends on the existence of social life and communication between human beings, each one being endowed with a brain/mind capacity. In this sense science depends on society; without social communication the very process of science, which is focused as much on consensus based validation as on systematic observation, would not be possible.

What is Science's Aim?

The aim of scientific activity is to take some part of reality and disaggregate it—that is to say, to break it down to its elements and focus on how they are related to each other. Typically, one is trying to understand a process, namely what causes the dynamism of an interaction, is such exists. In scientific language this kind of study is called research. In order to carry out such research one must undertake a number of tasks: one must define, delineate and demarcate the subject one is going to investigate and one must determine the methodology and the tools or techniques to be used. The subject of investigation might be a socially pressing problem, such as finding the causes of a disease; or simply something the scientist is curious to find out, such as some special astro-physical conditions in outer space, or on a broader scale the very nature of our universe. Scientific methodology refers to the strategy and the principles adopted in focusing on the question posited by the scientist. For this purpose certain concepts will be operationalized and hypotheses will be set up; the latter usually deriving from existing theories. The researcher's art is in identifying the question that can be addressed and which will uncover revealing evidence with theoretical implications when formulated as a series by specific hypothetical predictions to be tested. The outcome of such testing will either further support or begin to refute the logic of the theory from which the hypothesis took expression. As for tools, these may include sophisticated devices for observing reality, such as powerful microscopes, entire laboratories, spacecraft, or the gathering of statistical data through surveys and their analysis by means of computers; or again qualitative methods such as in-depth interviewing by psychologists or participant observation as engaged in by anthropologists.

All the above activities of the scientist boils down, in the first instance, to observations of reality, which produce information that the scientist will interpret, in order to provide explanations of whatever exists and happens in a more limited consensual reality. This means that since our observations are based on sense-impressions or (as some call it) sense-perceptions, and since according to many scholars there are processes involving physical as well as mental 'states', processes which we do not fully understand, the very source of science, like that of all human existence, remains a mystery. It may be quite convincing to say that so long as this basic process works, it is of no consequence for scientific results whether we shall ever uncover the nature of its source. However, there are some implications for science arising out of our relative ignorance in this field. It is necessary, therefore, to consider the extent of our knowledge of what perception is and to highlight the implications for science.

All about Perception

It is clear that we pick up information through our senses all the time. So does the scientist. What are these sense-data? They are clearly not substances or events but rather our knowledge of them—which is both partial and interpreted. They are, however, related to material substances and events, in so far as they represent them to us, but what exactly is their nature is a matter for metaphysics (an abstract philosophical consideration). Thus, we see a dazzling flash of light which we call lightning caused by a discharge of electricity between clouds, or the scientist observes under his powerful microscope the behavior of some cells when viruses attach themselves to these cells. These are the sense-data which scientists perceive. As scientists we are conscious of their existence, we can organize them in categories, see their effects on one another, search for causal connections, and so on. Examples may be the observing of certain interactions between chemical substances or the interaction between human beings. In both cases the scientist puts the elements involved and their interactions into 'types' or 'categories', and he will try to understand how the interacting elements affect each other, and what caused the interaction. In undertaking scientific work one is constantly engaged in perception, even when reading the computer output in the form of numerical results accruing from research. (Price, 1954; Rosenthal, 1971; Broad, 1960).

The process of perception is, therefore, crucial in scientific work. Nowadays, the most widely accepted explanation of how this process works is the physicalist or purely materialist explanation of perception. It runs something as follows: Sense organs receive, through highly specialized sensory cells, signals which are transferred to nerve endings; that is to say, the specialized cells react chemically or mechanically to the surrounding physical environment. For instance, smell is a reaction to some chemical compounds; sound waves, on the other hand, are picked up mechanically by very fine hairs. A more elaborate example would be sight: here the light which may be defined as photons or packets of electromagnetic energy, traveling as waves, interact with the chemical molecule in the eye, i.e., the pigment. The latter, as a result changes its three dimensional confirmation (shape), which triggers a chain reaction in the specialized eye cells, so that a transfer of information takes place by alternate chemical and electronic neuro-transmissions through the synapses, and so on from one cell to another, until the brain cells are reached. Here a sophisticated processing of the information takes place, the result of which is vision. This brain process is as yet not fully understood. According to John Ziman "The fate, within the brain, of the signals leaving the optic nerve is little understood" (Ziman, 1978).

The brain, being the most complex organ of the human body—a kind of super and extra fine computer with a phenomenally high capacity for the most intricate and precise processing of information—is the key to perception, as it is to other human activities, such as thinking, feeling, imagining and remembering. The question is whether there is an *identity* between such mental states and brain processes, or alternatively whether there are certain special properties, connected indeed with brain and neuro-physiological processes, but which are properties that create an extra human dimension called *the mind*. The latter view leads to a dualistic position, which would seek to understand the workings of our intellect as such, in its own terms, while not denying its connection with that different materialist side, that is the brain and the entire neuro-physiological complex.

The main argument in favor of a dualistic view is that each individual has the ability of introspection, that is of looking into one's mind, so to speak. It is this 'privileged access' to one's own mind, this privacy, which makes possible mental states such as, for instance, being in a state of moral indignation. It may be argued that it is this same extra dimension of the human intellect which generates the special intuition and insight that a person may have, abilities which are, like perception, crucial in scientific work. Of course, those who disagree with the dualistic position hold that such finely-tuned mental activities as insight may, in the final analysis, be reducible to the highly complex patterns of brain imprints and code-associative processes taking place in what is basically a biological network.

These interesting questions, far from resolved, have produced a vast literature, especially in philosophy and cognitive psychology, which I will not pursue here. From our point of view, however, we shall see that perception and thought are the two pivots upon which scientific work rests. This has certain implications for the twin claims of classical science: rationality and objectivity.

Is Objectivity in Science Possible?

A hundred years ago Karl Pearson, the British scientist, described "the scientific frame of mind" as one which eliminates individual bias, in the sense that judgments of such a mind are independent of the idiosyncrasies of the individual. He saw modern science as "training the mind to an exact and impartial analysis of facts" (Pearson, 1937). It is the capacity that human beings have to reason, to use logic, that produces scientific laws. Thus, according to Pearson, the scientist uncovers the laws of nature by projecting his perceptive faculty and his powers of reasoning onto the multitude of phenomena he examines.

The confidence Pearson expresses in the ability of the scientist to be rational and objective[1] echoes in many ways the views of Francis Bacon in whose writings, more than two-and-a-half centuries before Pearson, we find a scheme of how to put man in command over nature, which Bacon held to be "the grand object of all science". Bacon recognized the fallacies to which the human mind was prone due to tendencies to accept preconceived ideas, subjective ways of looking at nature, the influence of existing language, and the uncritical following of past philosophical systems and erroneous scientific methods. Hence he proposed a new, in his view foolproof, method for science: one which "insisted on dismissing prejudices and misconceptions of all kinds and on the close and methodical observation of facts". His empirical method relied on induction, that is to say the examination of all relevant details regarding a given phenomenon, which one tries to abstract into a general scientific statement/law. This method, he thought, when applied properly by any scientist would lead him (the scientist) automatically to an objective truth about the essence and nature of the physical world. One could thus accumulate more and more correct knowledge about the world, and this would entail the progress of humanity toward universal truth (*Encyclopaedia Britannica*, 1957, 1976).

For centuries, right up to our own times, the idea of Bacon that the scientist is basically an objective observer who casts aside all presuppositions and who adheres to exactness and never relies on guesswork, was held to be correct. In a similar vein, his contemporary, René Descartes argued that in order to achieve scientific certainty all existing ideas had to be doubted, namely that the scientist had to rid himself of all the accumulated preconceptions. Modern philosophers of science and modern scientists are denying the possibility of achieving these aims. Scientific objectivity and certainty, according to them, is impossible. To start with, there can be no neutral description of the world around us; neither can there be absolutely correct description.

The reasons for such negations are philosophically and sociologically easy to uphold: (a) we can never have a clean slate, a start from scratch, that is to say, our observations can never be presuppositionless; (b) even the scientist's insistence on standards of rationality and objectivity are agreed conventions and hence they are the outcome of human interaction and experience (Prado, 1987; Churchland, 1992); (c) without some basic ideas we could not compare or distinguish between things and we could not decide on their relevance to our subject of research (O'Hear, 1989).

But the negations of certainty and objectivity become even more convincing when we consider the scientist's own work. Developments in atomic physics finally shattered the solid basis that science was supposed to rest on. Werner Heisenberg, the famous German atomic physicist, provides us with an account of developments in atomic physics; an account which has been quoted over and over again because of its clarity and because it has revolutionized our view of science.

Very briefly, this is what Heisenberg says: When we try to observe the tiniest particles of matter, the elements of which atoms are made up, we cannot avoid interfering with these elements and their behavior through the physical process of observation. That is to say, the instruments scientists use affect these tiny particles, and, to say the least, the scientist cannot be certain that in his process of observation he does not disturb the thing observed. A similar view has been expressed by scholars who have considered the relations between the act of perception and reality. The problem they have highlighted is that the brain, the mind and the whole nervous system which make the process of perception possible, and the process itself, are all part of the reality in which perception takes place, including the phenomena perceived; therefore, not only does reality act on perception, but the process of perception can also alter that reality. To quote H.H. Price: "Even if there be something which is given it is quite impossible for us to know it. For if we attempt to describe any so-called datum, e.g. this view which I now see, the very act of describing alters it" (Price, 1950). Returning to the example from atomic physics, Heisenberg states that there is no point in asking whether the particles exist in space and time objectively, since all we can say is that given the interplay between our observational instruments and the particles observed, we have a certain knowledge of nature and its elements. However, Heisenberg and other physicists, like Niels Bohr, maintain that this knowledge does not give us absolute certainty about the accuracy of our observations. The question has even been raised whether the physicist can actually *see* a certain subatomic particle when he points to scratches on photographic negatives which he claims to be its tracks (Gregory, 1981). Consequently, scientists have become conscious of the limitations of modern science and cannot uphold a view of science which claims objectivity and certainty as its basic tenets (Heisenberg, 1958; Bohr, 1958).

The limitations we just mentioned do not detract, however, from the scientific achievements of our age. Space technology and laser surgery, for example, are testimony to the fantastic achievements of modern science. What is it, then, that enables science to advance? What is its force? Can science progress to the point of uncovering the ultimately true natural laws that govern our existence, our world, indeed the universe and its inception?

Progress in Science

A pragmatic definition of what is meant by progress in science would stress two achievements of scientific work: (a) science has been able to find out increasingly more about "what the world is really like"; (b) science has helped humanity to cope better with reality (Prado, 1987). This means that not only does science add to and improve upon our knowledge of what is happening around us, but it also enables us to deal with our lives and our environment in a more efficient and beneficial way to us. We shall overlook for the moment the environmental hazards created by the reckless application of science and technology in our modern civilization. The point I am making is that in principle the progress of science in a society can be assessed in a practical way. Thus, modern science has enabled a schoolchild to have more valid, reliable and useful knowledge about the physical world than Newton had, thanks to the scientific advances made by Rutherford and Einstein; and the applications of medical science has enabled us to deal with infectious diseases much better than people were able to deal with such diseases a hundred years ago, and to traverse distances in a fraction of the time which was required of previous generations (Prado, 1987).

Science both accumulates knowledge and advances it. It advances by providing more accurate descriptions, better explanations and more successful predictions regarding our lives and our world around us. This is not just a result of the incremental improvements due to the accumulation and organization of facts. To understand how science advances one must refer to the principle underlying the scientific enterprise. This may be summed up in one sentence: *you learn from past accumulated knowledge but your aim is to break out of it*. To break out of accepted ideas is to produce radically new ideas and thus to improve your knowledge. This is the strength of science and herein lies the difficulty of the task.

Two well-known philosophers of science provide us with different views regarding this aspect of the scientific enterprise. On the one hand Thomas Kuhn maintains that scientists do not carry out their research on this principle. On the contrary, they work within certain frameworks provided by existing theories, assumptions, research practices and standards. Kuhn calls such frameworks 'paradigms', and shows that for long periods of time scientists accept existing paradigms uncritically, and take them for granted. When they come across observations which appear to be contrary to the ruling paradigms, they labor hard to explain away such contradictions as particular anomalies caused by special conditions, rather than accepting the contrary evidence as indicative of a deeply embedded flaw and thus destroying the credibility of the whole framework. Karl Popper, on the other hand, insists that the scientist is aware of the impossibility of ever fully proving the veracity of a scientific theory. The scientist may at any moment come across new facts which would falsify the accepted explanation of a phenomenon. Faced with contrary evidence, according to Popper, the scientist is forced to search for an alternative explanation and thus abandon the former theory. This means that the scientist constantly tests existing theories, which, so long as they are not falsified, can be regarded as corroborated for the time being, but never fully proved. By contrast, Kuhn portrays scientists as more conservative and group oriented. The consensus does not fail due to a single falsification event, and the community of scientists does not abandon an explanatory framework until a better, or at least a more promising account of the known and anomalous events has been offered as a new paradigm contender. Until this happens a field is in "crisis", and coming out of that state may require a "revolution". However, Kuhn's point is that a crisis of confidence in a paradigm does not mean a denial of known and accepted facts which any new paradigm must still explain in a satisfactory manner.

The following are two illustrations, quoted from John Kemeny, concerning this discussion. The first example is from the natural sciences. Scientists once believed that the process of burning consisted of a certain material substance, called phlogiston, being discarded. Doubts arose regarding this theory, since, if it was true, the weight of the burning material should be reduced. The scientist A.L. Lavoisier showed, however, that the opposite was true, that is that a burning material gains in weight. The existing theory was, therefore, untenable in his view. He further claimed that there was no evidence for the existence of phlogiston and that it was not needed to explain any known chemical reaction. A new theory was posited, which held that burning was possible because a substance was taken in during the process of combustion. This substance was the oxygen in the air. The new theory can be corroborated by a simple experiment. The burning will cease in a closed space in which fresh air is not replenished (Kemeny, 1959). Lavoisier was credited with the "discovery" of oxygen and by propounding an explanation of combustion focusing on the role of oxygen in that process, the phlogiston paradigm was displaced.

There are two possible interpretations for this outcome. For Popper it is clear that scientists could not reconcile the two frameworks and, therefore, chose "rationally" the simpler and better supported account and abandoned the phlogiston framework as a scientific "dead end". Those who tried to save the old tradition were no longer true scientists, but nonobjective dogmatists. Kuhn, on the other hand, emphasizes that everything is relative to the paradigm which is accepted in science. He regards those who tenaciously tried to reconcile the new evidence with the old framework as still being acceptable as scientists. While both explanations are possible, Kuhn's account leads to a situation that ultimately leaves no rational ground for choosing between paradigms. In contrast, Popper's interpretation is that scientists' allegiance to a prevailing theoretical framework is determined rationally on the grounds of evidence, utility and predictive power assessed by the scientific testing of theories.

The example just mentioned and many other experiments seem to support Popper's idea about how science proceeds. He deems the scientific process of testing existing theories so crucial that he draws a demarcation line between theories which can be tested and belong, therefore, to the realm of true science, and theories which cannot be tested and he regards them, therefore, metaphysical or *non-scientific*. A case in point might be the Marxist theory of the class struggle and its historical prediction of the ultimate attainment of a classless society. Since we have no way of knowing whether and when we have reached the final, ultimate, historical stage of a classless society, since different interpretations are possible as to a society being classless or class-divided, such a theory cannot be tested rigorously. As to the socialist revolutions that have occurred, and events and structures of societies they have led to, these provided debatable evidence; and discussions surrounding such events have proved inconclusive thus far.

Popper's demarcation between true science, such as physics and other natural sciences on the one hand, and non-science, such as the social sciences and history on the other hand, is an extreme position. Experimentation and testing *is* often possible in the social sciences, albeit not always in the rigorous manner of the natural science experiment. This is where we introduce the second illustration of an experiment which is basically anthropological, yet it brought about reference to other disciplines including the natural sciences, in an attempt to test a theory. The illustration will show the complexity of testing theories concerned with social and historical developments.

The case is that of the origins of the Polynesian population and culture. Certain cultural similarities between the South Sea islanders and the inhabitants of areas in South America led to the theory that the origins of Polynesia were to be found in the more distant America rather than the nearer Asian areas. But there were doubts about such a possibility on account of the belief that pre-historic America did not possess adequate craft and the necessary navigational skills. To test the theory, Thor Heyerdahl, the Norwegian ethnolcgist, undertook a scientific expedition and sailed, soon after the Second World War, in a primitive raft, called "Kon-Tiki", until he reached the Polynesian Pacific area (Heyerdahl, 1963). As Kemeny puts it: Heyerdahl and his companions "risked their lives to test this theory" and "they found out a great deal not known before. They found a plentiful supply of palatable fish in these unknown waters, they found that the looseness of the raft prevented flooding, and they found out that, while steering was impossible, the prevailing currents carried them unerringly to their destination. Over a hundred days later they arrived on these islands and were greeted by the natives who related their old legend according to which the great god Kon-Tiki had brought their ancestors to the islands on just such a craft. Thus they verified their theory (which previously was in general disrepute) and completed a thrilling chapter from the history of the application of the scientific method" (Kemeny, 1959).

Attractive as the above story may be, one could not say that it was sufficient to "uphold a thesis that contact between prehistoric Polynesia and America had been made". The outcome of the expedition, however, triggered further attempts to corroborate the theory, by means of adducing linguistic, genetic and archaeological evidence. While the linguistic and archaeological findings were not conclusive, the biological approach, studying the blood groups and gene frequencies, did show that there *was* "a blood genetic relationship between American Indians and Polynesians" (Ferdon, 1968). Thus, testing the theory from several angles has given it credence and has added to our knowledge concerning human population migrations and cultural diffusion.

Are there Natural Laws?

It would appear that progress in science means that there is a direction that science takes towards the discovery of the laws of nature, namely that at rock bottom there are certain natural rules that govern our existence and that progressively science uncovers for us these rules. What we have to do as scientists is to observe all that we are aware of, to think rationally about the connections between the phenomena we perceived, to formulate certain hypotheses about these connections, and to test them experimentally. In this way we come upon the basic rules about the motion of the planets, the laws of gravitation, the basic biological process of the circulation of blood, the instinct of gregariousness among animals, and so on. It is through our perception and intellect that we formulate our ideas about what happens in nature. It could be argued, therefore, that the scientist manages to represent the laws of nature in the form of a consistent exposition. That is to say, that scientific laws reflect the natural laws which have come to light and which in a way exist independently and prior to scientific formulations. If this is so, then one could also hold that scientific laws, so long as they faithfully reflect the laws of nature, have universal validity. This implies that what exists or happens or, indeed, even that which we see as changing is determined by the immutable laws of nature. This is the determinist view in science.

The issues concerning natural laws and scientific laws are more complex than our presentation so far suggests. For one thing, scientific laws change but natural laws, by definition, do not. In other words, to say that natural laws exist is to assert that there is something absolute and immutable in reality, and that these immutables are there irrespective of whether human life or science exist or do not exist. But human life and science do exist and they are related to reality insofar as human perception picks up facts and the human intellect coordinates the observed facts in order to produce scientific explanations of that reality. Moreover, many of these scientific formulations when applied to reality seem to work very well, such as drugs curing illnesses, yet some have failed and others have been improved so that their application has become more successful.

Scientists and philosophers over the last hundred years, Karl Pearson, Bertrand Russell and Gilbert Ryle to name just a few, are all agreed that scientific laws are mental constructs, the projection of human reason into phenomena, the products of an ordering intellect applying logic to phenomena perceived, and always temporary—subject to change. But, as an example, this does not contradict the idea that gravitation is a universal force and that its existence in nature is absolute, namely it is there whatever we may think or do; it is inescapable. But as expressed and explained by our scientific law it is just one way in which the human intellect understands it. Furthermore, the understanding of the law of gravity has changed over the years, from gravity understood within the framework of classical physics which regarded the universe as static, to gravity understood within the framework of modern physics which sees the universe as expanding. The same nature and the same force in it can be given to different kinds of understanding and explanation in science (Hawking, 1989).

Natural law is, therefore, something nebulous; something that may exist out there in the distance but we are not certain what it is; and if it does exist, we assume that it affects us constantly. Once again, as mentioned above, according to this view we are facing a determinist existence, since all that happens is due to the existing natural rules and all the facts in the universe and all our doings, including our future, are predetermined by those rules. However, determinism need not be seen in such absolute terms. There may be certain ground rules and certain boundaries that put limitations on the occurrence of phenomena and on the actions of people, but this does not rule out the possibility of alternative occurrences or the choice of actions. As Gilbert Ryle put it: "Laws of nature are not fiats". As an example of natural occurrence one could give the atmospheric conditions which produce different possibilities of wind and cloud formations and predict their chance of resulting in rain over certain areas and not over other areas in the same region. As for human choice one can mention Ryle's illustration of a game of chess which is governed by agreed rules, without the latter determining the actual moves of the contestants (Smart, 1971).

Finally, developments in modern physics since the earlier part of this century have introduced the idea of indeterminacy in natural events, which does not see natural processes as entirely regular. As a result, recourse had to be made to the rules of statistical probability, and 'chance' had to be given a more prominent place in a science which could no longer distinguish itself as an *exact* science. In any event science has many faces and it produces different kinds of knowledge, various levels of understanding, and varying degrees of predictability. So it is that while the science of astronomy can tell us the exact moment when the next eclipse of the moon will take place, geologists have difficulty in predicting with such accuracy where and especially when the next earthquake will take place, and what degree of power it will register on the Richter scale.[2] Neither have scientists fully elucidated nor have technologists overcome the problem of metal fatigue[1] or the sudden malfunctioning of electronic systems. Again, in demography, for instance, pretty accurate average life-expectancy can be calculated for an entire population, barring wars or extraordinary events such as natural disasters, but no such accuracy can be achieved for predicting the longevity of any individual. It is not at all surprising that the questions of stability and regularity as against unpredictability and even chaos concerning the natural sciences, and the lack of success in problem solving or in predicting events in the social sciences, have taken the center of interest in the 1980s and the 1990s. It is to these questions that we turn to in the next chapters.

NOTES

1 Despite such proclamations, Pearson himself exhibited subjectivity and prejudice in dealing with the question of race and disseminated through his journal, the *Annals of Eugenics*, unsubstantiated and nonsensical ideas disguised as scientific theory (Barkan, 1992).

2 For improved possibilities of prediction in theses areas, but also for continuing uncertainties, see the detailed examples of John Casti and Michael Scriven (Casti, 1993; Scriven, 1988).

CHAPTER TWO

MAKING SENSE OF REALITY

I stated in the previous chapter that science deals with reality. Our interest now turns to two basic questions concerning reality. First, what basic feature or features characterize reality? Is reality orderly, stable, structured in a certain way, changing and developing in an orderly fashion; or conversely, is its existence based on randomness, with events occurring one way or another, exhibiting much chaos, disintegration and instability? Or again, is it possible that reality contains both features: chaos and order? And if so, how would this be possible? The second question arises out of the first since it, too, is a kind of feature of reality and concerns something very basic about reality. This is the question of boundaries or limits which are thought to encapsulate reality and the effect such limiting conditions have on that encapsulated reality: for example, what are the biological boundaries or conditions within which cells can exist and multiply; or into which domains do galaxies expand?

Reality as a Constant Process

The first thing that scientists, as well as philosophers of science, do in their attempt to understand reality and its basic features is to propose conceptually what reality is. In the way of ancient thinkers like Heraclitus, or modern scientists of this century like A.N. Whitehead, the current fashion in science is to conceptualize reality as a process. One can visualize, therefore, reality as being in constant motion: the planets move along their orbits, stars collapse into black holes, cells multiply and move along trajectories to different positions, living organisms develop and die, clouds form and disperse, individuals' psychological states come and go, societies change slowly or violently, and so on endlessly. As David Bohm puts it: "Not only is everything changing, but all *is* flux" (Bohm, 1980).

An excellent example of the constant process of motion in nature can be provided by the history of scientific theories concerning heat. During much of the 19th century scientists predominantly accepted the caloric theory of the transmission of heat, despite some evidence against it. It was commonly believed that heat was simply a substance called "caloric", which was described as a "subtle fluid" capable of moving around more freely than any other matter and could, therefore, penetrate any other matter rapidly and with ease. The early evidence against this theory was mainly provided by the experiments of Benjamin Thompson (Count Rumford; 1753–1814) at the end of the 18th century, who attempted to prove the idea of the Dutch chemist Boerhaave, that "heat is a vibration at a frequency which is too high to be heard as sound but which is sensed by the skin as hotness". His experiments showed that heat had no weight, hence it could not be some additional fluid; that heat can be produced by mechanical friction; that the atoms of matter are in constant motion, even in substances which appear to the eye as if at rest; and that the "peculiar and continued motion common to all liquids is caused by the instability or lack of equilibrium existing among their molecules". He also "measured, with fair accuracy, the relation between heat and work". Rumford thus "anticipated the kinetic theory of heat" (Stevenson and Byerly, 1995).

However, it was only in the latter part of the 19th century that the kinetic theory was established by Rudolf Clausius (1822–1888), James Clerk Maxwell (1831–1879) and Ludwig Boltzman (1844–1906). The theory is based on the fundamental conception that heat is produced by a chaotic motion of molecules. These constantly collide with each other just like billiard balls; a natural movement among molecules which will be found in all kinds of matter: gaseous, liquid, or solid, with a descending capacity for movement—for instance in solid material the molecules being so densely packed that there is little room for such motion. Although the theory applies to solids, liquids and gases alike, it was originally stated in connection with gases. The explanations of the theory developed by Maxwell and Boltzman have an essentially statistical character. The heat which is proportional to the kinetic energy of the molecules varies over a wide range, since their motion varies in the continually changing collisions among the molecules and the constant collisions with the walls of the container of the gas. Yet, because of the vast number of molecules involved, an average kinetic energy will be found. This remains the same, so long as no outside heat is added, and the uniformity corresponds with the observed fact that the temperature of the gas becomes the same throughout the container: there is thermal equilibrium. But as Maxwell himself held, if we took the endlessly long run in terms of chance and time, the actual motions of the molecules may not necessarily bring about the uniform temperature in the container. Yet in our experience we never come across such lack of uniformity and there is the highest probability that this will not happen. Nevertheless, we can say that: "the kinetic theory is essentially bound up with the conception that the apparently inevitable laws of nature are actually no more than laws of probability" (Singer, 1959).

Another modern scientific theory was anticipated by Rumford. He came close to the modern wave theory of motion when he referred to "undulations in the ether" (Goldstein and Goldstein, 1985). However, it was only at the beginning of this century that the famous scientists Max Planck and Albert Einstein provided convincing explanations for heat radiation and light transmission. Basically there is no difference between the two: radiant heat is a form of light which is invisible to the eye, while light transmission is electromagnetic radiation that *can* be detected by the human eye. Yet, both are "transmissions of energy between material bodies at a distance", and in both cases the transmission is in the form of waves similar to the ongoing oscillations in magnetic and electric fields (Bohr, 1958; Hawking, 1988; Goldstein and Goldstein, 1985). It is beyond our expertise, and it is not our purpose at all, to explain scientifically the phenomenon of energy transmissions, but rather, by means of the above examples, to point to reality as a constant process.

What Does Reality Consist of?

To the layman this seems a strange question. After all, we can see what reality consists of: the universe with its stars, our world of nature, humans and their artifacts, and so on. However, the layman will be equally surprised to hear a physicist or to read the works of a scientist who maintains that all those things mentioned, whether it is the sun, water in a pond, the air we breathe and the light we see, indeed our own bodies, are made up of some ultimate tiny substances and forces: atoms and their tiny components interacting through electromagnetic forces in the form of waves, thus forming the somewhat bigger building blocks of molecules, leading to the structures which appear in different forms of physical matter, plant and animal life.

Developments in modern science produced a number of fascinating discoveries regarding the tiny substances and basic forces of which reality is made up. Atoms were once thought to represent the smallest elementary particles. However, between the 1890s and the 1930s three British physicists, J.J. Thomson (1856–1940), E. Rutherford (1871–1937) and J. Chadwick (1891–1973), who were all associated with the Cavendish Laboratory in Cambridge, showed that atoms were divisible and have an internal structure containing elements: electrons, protons and neutrons. Then came further discoveries of even tinier elements which scientists call quarks, giving rise to the question whether science will ever discover the *ultimate substances* of which reality consists. These tiny particles are charged with energies which are in fact waves. It is the wavelength encountered in experiments which tells the scientist how high is the energy of the tiny particle: "the higher the energy of a particle, the smaller the wavelength of the corresponding waves". So, the higher the energy discovered by encountering smaller and smaller wavelengths the tinier the particle. As Stephen Hawking tells us: "And so we know that particles that were thought to be 'elementary' twenty years ago, in fact, were made up of smaller particles. May these, as we go to still higher energies, in turn be found to be made from still smaller particles? This is certainly possible, but we do have some theoretical reasons for believing that we have, or are very near to, a knowledge of the ultimate building blocks of nature" (Hawking, 1988). This statement suggests strongly that scientists do not know what are the *ultimate* substances and forces in nature.

The idea of David Bohm of regarding the universe as "an undivided and unbroken whole" which is characterized by constant movement, through interactions and transformations, helps us understand why one must deny the supposition that science will ever reveal the ultimate substances of reality. To back up his argument he points to the fact that the 'elementary particles', which *are* known to exist, can be created, annihilated and transformed, which indicates that even these cannot be ultimate substances but, rather, "relatively constant forms, abstracted from some deeper level of movement". The analysis of such deeper levels may uncover even finer particles and further properties of such particles or collections of particles; yet these will remain only parts and aspects of "an unknown and undefinable totality of flowing movement" (Bohm, 1980).

What is Entropy?

One of the best examples of this constant flowing movement is the transmission of energy between the various entities in the universe. The dynamics of life on earth as well as the workings of the entire universe are based on such transmissions of energy, energy being defined as "the capacity of doing work". In simplest terms, my moving a chair involves my expending some energy. I replenish this energy by the intake of nutrition. Human beings produce food by expending energy. In the industrial-technological world this means the use of material resources and sources of energy, such as metals for producing machines and oil for using them; or the harnessing of the energy given out by the sun to heat water, or the ocean waves used to drive waterwheels. Can this cycle go on for ever? Or is there a danger that such sources of energy could eventually be exhausted?

There is much dispute around this question, the optimists arguing that, despite the depletion of ancient forests and the rapid diminution of supplies from existing oil fields, there are still potential sources of energy on our planet and in the universe. In the 20th century science added the new sources of nuclear energy, and maybe one day science and technology will enable mankind to harness and utilize energy from distant stars! There is another kind of question, however, that arises from the simple example I have given above. Has the energy I used in moving my chair been wasted? In so far as it has achieved the purpose of moving the chair from one position to another, the energy put in has not been "wasted". But in the process of this vectorial movement a good deal of energy has been dissipated. Quite apart from the net energy needed to move the chair, depending on the size and weight of the chair, the frictions involved as a result of obstacles, the distance to which it was moved, and the velocity of the act of moving it, the gross energy used was even greater. This is so because the total calories I used in this act has been available to me due to nutrition. But in the very process of nutrition, using my digestive system including the necessary breathing and other biological functions, involved the using up of energy. Furthermore, the human body is a complex and mostly efficient machine but not a perfect one. In its acquiring and spending of energy a fair amount of it is dissipated due to different conditions and events occurring in the body and its environment, during the very process of transferring into it or out of it quantities of energy.

It would appear, from the above example, that energy is not simply transferred and that a part of it will cease to be available for doing work, since there is always some wastage in the process of its use. Scientists, especially since the middle of the 19th century, have considered the question of the conservation and dissipation of energy, which led them to the theory called entropy. Basically, energy may be regarded as a force at work. It exists in different forms: kinetic energy is at work—for example, where a material object, due to gravitation, falls through the air and exerts force on its way; electric currents conduct energy to our electric appliances; and thermal energy is at work when we burn fuel in an internal combustion engine, so that our car will move. In the last example we can see how engines can convert one form of energy (thermal or electric) into another form of mechanical motion (kinetic). In classical science it was held that energy conservation is possible where an ideal machine simply converts energy without any loss. Scientists in the 19th century pointed out, however, that in real machines there is not simply an exchange of energy from one type to another, but rather that internally the system of the machine uses up energy due to friction or heat losses, which are irreversible processes. Therefore, we must distinguish between "useful" exchanges of energy and "dissipated" energy that is irreversibly wasted (Prigogine and Stengers, 1984).

Let us now look more closely at the idea of entropy that arose in connection with the two basic principles of thermodynamics, a branch of the natural sciences that developed in the 19th century. Thermodynamics is the study of processes involving heat changes and the availability of energy to do work. Experiments in the 1840s of the British scientist J.P. Joule (1818–1889) established that heat and work were convertible—that is, one can transform mechanical energy into heat and heat into mechanical energy. This was the basis of the first 'law': that energy, although it can be changed in its form, is conserved in that it cannot be created or destroyed—it is all there in its existing sources. However, the heat drawn from a source can be converted into work only by some of it being 'wasted', although not 'annihilated'. According to the second 'law' there is a preferred direction in the process of heat transfer—as expressed by the German physicist Rudolf Clausius (1822–1888), "heat cannot by itself pass from a colder to a hotter body"; of its own accord it will do the opposite, that is it will transfer/flow from a hotter to a colder body.

But more importantly, Clausius put forward in 1850 the theory that in the process of the conversion of heat into work only part of the heat supply can be converted into actual work. In this connection Clausius introduced the idea of *entropy*, which is a measure of the transformability of heat, that is the extent to which existing energy can be converted into work. The word *entropy*, which he coined, is from the Greek meaning "turning into". The most significant point in his theory is that all natural processes involve an increase in entropy. This is so because heat transfers only when there is a difference in temperature and, as we said, the direction is from hot to cold. In this process some of the heat is 'lost', that is: it is no longer 'available' to do work. Thus "every transformation of heat to work...is unavoidably accompanied by the transformation of another quantity of heat from an available to an unavailable form". Since all spontaneous processes are irreversible, this means that all these processes are accompanied by an increase in entropy, which in an isolated system reaches the maximum when the system is at equilibrium. This is the peculiar nature of heat-energy, in that it is always losing from its transformability; in other words: entropy, the dissipation of available energy, is unavoidable. Clausius, therefore, summed up the principles of thermodynamics by saying that while the energy of the universe is constant the entropy of the universe tends to a maximum (Mott-Smith, 1964; Zernike, 1972; *Encyclopaedia Britannica*, 1992).

The idea of entropy was supported by another important scientist of the 19th century, William Thomson (Lord Kelvin), who expressed it as a continual degradation of energy (Asimov, 1966) Entropy has also been expressed as "a measure of the degeneracy of a macroscopic state" (Eisenberg and Crothers, 1979). Yet another way of applying the idea of entropy is to show how useful energy is lost where certain spontaneous processes change a situation of existing order to one of instability and disorder. Some energy is then used up so as to bring about a new equilibrium situation, a new order. The example usually given is that of contact and transmission of heat between two materials of different temperatures. Our initial state of two materials, each with its own temperature, signifies an existing order. When in contact the unbalanced distribution of temperature (a less ordered state) will tend to a uniform temperature between the materials. But the achieving of this new order or equilibrium will have involved some entropy, that is to say there is some loss of useful energy. An uncontrolled fire which dissipates into the atmosphere is an even better example of the loss of order and an increase in entropy.

The principle of entropy and (its corollary) of irreversibility have important cosmological implications. Logically, it is difficult to accept the notion that, through dissipation, energy is totally lost. After all, the heat from the uncontrolled fire mixes into the atmosphere; it does not escape or disappear entirely; it is still somewhere in the universe.[1] But it is true to say that the heat molecules from the fire, which have dispersed into the atmosphere, cannot be brought back in an organized form. This means that while total energy may still be the same in the universe, the energy available to do useful work will have decreased. And since such spontaneous processes are at work all the time, it is correct to say that entropy is continually increasing. To put it another way, the universe may be said to be "running down" (d'Abro, 1950; *Encyclopaedia Britannica, Micropaedia*, 1992).

Order and Chaos

The most significant aspect of the principle of entropy is the direction of physical reality toward deterioration; namely, we are passing from a more highly ordered reality to a less ordered or disordered one. Ilya Prigogine grapples with this problem when he highlights the two apparently opposing theories in science: entropy and evolution. He argues that entropy does not necessarily entail a deterioration, but rather a reorganization of reality. And since the reorganization is evolutionary, leading to more complexity, there is no basic contradiction between the two principles.

Prigogine's main arguments are as follows: If we look at organic, rather than inorganic, reality we shall find in the spheres of biology and chemistry that living organisms and chemical matter operate as open systems. These open systems exhibit continuous processes of change, they react to their environments, produce feedbacks, and above all are liable to strong fluctuations. Since systems have sub-systems, the fluctuations in one such sub-system may result in change throughout the system. Consider the human body as a system, with its blood circulation as a sub-system. A change in the latter, due, for instance, to a change in the oxygen level in the blood, will affect the entire system of the body.

Prigogine's examples are mainly in the micro-biological and chemical fields. His own studies and research conducted by others in these fields lead him to the conclusion that a fluctuation or a combination of fluctuations, especially when they push the system beyond some threshold, may be so powerful that they produce a revolutionary moment, which he calls a "bifurcation point". It is impossible to say in advance what direction such a basic change will take. It could bring about a disintegration of the system leading to complete chaos, or it could produce a leap to a higher order, a reorganization into a more complex system. In other words, he is dealing with non-linearity and dissipative structures, which are dynamic and are not directed towards homogeneity and equilibrium. Looked at macroscopically, such structures are holistically ordered, such as where a vast number of molecules interact and show synchronic changes—that is to say, an ordered totality. One of Prigogine's examples is "a chemical clock" which he illustrates thus: "Suppose we have two kinds of molecules, 'red' and 'blue'. Because of the chaotic motion of the molecules, we would expect that at a given moment we would have more red molecules, say, in the left part of a vessel. Then a bit later more blue molecules would appear and so on. The vessel would appear to us a 'violet', with occasional irregular flashes of red or blue. However, this is *not* what happens with a chemical clock: here the system is all blue, then it abruptly changes its color to red, then again to blue. Because all these changes occur at *regular* time intervals we have a coherent process!" And he continues: "Such a degree of order stemming from the activity of billions of molecules seems incredible, and indeed, if chemical clocks had not been observed, no one would believe that such a process is possible. To change color all at once, molecules must have a way to 'communicate'. The system has to act as a whole" (Prigogine and Stengers, 1984). The complete changes are obviously induced in some way, which Prigogine and Stengers interpret as a kind of 'communication' that affects molecules.

Prigogine distinguishes three states in which systems may be: equilibrium, near-equilibrium and far-from-equilibrium. The reasons for changes from one state to another are various; in the case of fluid flow, for instance, once a flow rate of flux is reached, turbulence may occur. But such turbulence is not to be identified as complete disorder, for turbulent motion that appears at the macroscopic scale as irregular or chaotic, hides the coherent behavior of millions of molecules seen at the microscopic scale. Furthermore, transitions from one state to another, from stability to chaos, and from disorder to order, show that much of the material of which reality is made is capable of spontaneous self-organization. Therefore, out of chaos, Prigogine tells us, "dissipative structures" of a higher order may come about. We see most clearly how self-organization works in cell biology, where non-linear feed-back and autocatalytic mechanisms, as in DNA[2], enable cells not only to multiply but also to pass on the genetic code (Prigogine and Stengers, 1984). It is in this capacity of self-organization, and in the possible changes in existing states of pre-existing order, that Prigogine sees the source of evolutionary development. In this sense the entropic process in reality, seen as a breakdown in order, does not contradict the evolutionary process, seen as reorganization into a higher order. However, we must remember that organisms exist as open systems, and it is, therefore, worth mentioning here Niels Bohr's remark that organisms are continually supplied with free energy (Bohr, 1958), for this seems to suggest that in the very long run entropy may bring us to the limit of reorganization and development in our life-world.

The upshot of the above discussion is that while reality appears to contain aspects of order, structure and stability, it also has a vitality expressed in processes of change, fluctuations, disorder and spontaneous reorganization and restructuring. Reality is, therefore, characterized by an openness, and by features of randomness and chance (a topic which we will take up in the next chapter), as well as by structural and systemic features. In science today we come across topics such as "the physics of disordered systems", which conjure up in our imagination a basic contradiction: if one talks about a 'system' then logically there must be some *order*: or to put it the other way, if it is *disorder* that prevails, is it still reasonable to talk about a 'system'? (Krausz, 1987).

Numerous explanations are given to such apparent contradictions by the study of chaos and the scientific understanding of its place in reality. James Gleick sees such studies as making up a new science. The reason why he regards the study of chaos as a new scientific approach is that it shatters traditional science's cosy assumptions about a basic stability and regularity in all that is connected with physical, chemical and biological systems and processes. The traditional scientist sees in disorder just a little 'noise' and wishes to disregard small non-linearities. The latter exist where the relationships between two or more variables are not strictly proportional. Take the variables of friction and velocity in the motion of some substance. There is not a constant importance to friction, since it depends on the speed with which a substance moves; on the other hand, speed depends on friction. At higher speeds friction has quite a different effect than at lower speeds, and changing degrees of friction will also affect speed differently. It is this changing relationship that we call non-linearity.

Non-linear relationships cannot be disregarded; the ideal scientific world where regularities can be separated from the disorder of experience is no longer valid. But this does not mean that the world is simply based on chaos and that the latter implies some basic instability. It is rather that our world, our reality is very complex, and what appears chaotic at one stage or on one scale transforms into stability at another stage or on another scale. The problem is that the various stages and scales form part of a larger system, so that "any system could have both stable and unstable behavior within it" (Gleick, 1993). To repeat the example above, a river is a flow composed of a mixture of regions of smooth flow and turbulent eddies. Irregularities in the terrain, requiring different fluid densities and different expenditure of energies in the flow, is part of the explanation for such a mixture. But even in a smoothly shaped pipe the flow becomes turbulent at a certain phase, say when the volume or speed of the flow increases greatly. Two points are worth noting: (a) on the larger scale of the entire flow the vortices represent the chaotic and irregular aspects of the flow, but on the smaller scale, within the tiny flows that make up the vortices themselves, the molecules behave uniformly and in a coherent fashion—it is their actions in unison that make up the vortices; (b) such descriptions in themselves do not explain at what point and why turbulence sets in: why is it at *some* speed that a fluid flow becomes non-smooth, or why at a particular distance form the source a smooth pattern of smoke emanating from a burning cigarette breaks up and becomes an apparently chaotic pattern of smoke?

The onset of turbulence, Gleick tells us, appears as a mysterious moment when an orderly system turns chaotic. Consider the following chain of action: "when a solid is heated, its molecules vibrate with the added energy. They push outward against their bonds, and force the substance to expand. The more heat, the more expansion. Yet at a certain temperature and pressure, the change becomes sudden and discontinuous" (Gleick, 1993). This means that there are critical points at which matter enters new realms, so that the molecules of the matter begin to behave differently. But the mere fact that unheated water, which is still, begins to bubble furiously when heated sufficiently, and that in yet another phase it evaporates, does not explain why these transitions may cause turbulence. The answer, albeit not a fully satisfactory answer, is that there are different "attractors", which I would define as "guiding forces", that attract the molecules of matter in certain conditions towards a particular pattern of behavior. On a larger scale, an example of two different "attractors" would be a stationary pendulum, its behavior being determined by a stability "attractor", as against a pendulum whose condition is changed by applying energy to it, thus producing the swings of the pendulum; the "attractor" or guiding force of the swinging pendulum is dynamic (Gleick, 1993).

In the case of the pendulum, the oscillations slow down as a result of loss of energy and tend towards zero without turbulence. But in other cases, such as telephone transmissions of sound, a change in the electronic frequencies, resulting from changes in the supply of current, can cause irregular noise, perceived by callers as annoying disturbance during a telephone conversation. The disturbance is due to "the conflicting pulls of a higher or lower frequency" (Gleick,

1993). Turbulence can, of course, have much more disastrous results, unless controlled or harnessed as in the burning of fuel in an engine. When turbulence is out of control, such as turbulent airflow, it can result in a plane crash. To sum up, in the words of James Gleick, turbulence "is a mess of disorder at all scales...It is unstable: It is highly dissipative, meaning that turbulence drains energy...It is motion turned random" (Gleick, 1993).

Some systems cannot function if turbulence occurs; they cease to exist if randomness takes over. They can exist only where regularity and rhythm dominate. A case in point is the biological system of the human or other animal organism. The human organism can be characterized as a system in "steady state". This means that the organism has a constancy: although matter/energies cross its boundaries, since the input of energy/matter equals the output, a kind of dynamic balance exists—homeostasis: a tendency of the system to maintain internal stability by the proper coordinated response of its parts to situations of stimuli which might disturb its normal condition and functioning. An imbalance in the ratio of white to red corpuscles of the blood, a disease known by the medical term of leucocythemia, produces death. Similarly, the irregularity of the heart beat known as fibrillation, namely the arrhythmic, uncoordinated contraction of the muscular chambers (ventricles) of the heart, controlling the circulatory system of the blood, can cause complete heart failure and death.

The very appearance of disease and death through lack of balance or rhythm, or the destruction caused in nature by radiation induced through atomic weapons, attest to chaos being part of reality. Yet the ultimate picture we get is one of a mixture of chaos and regularity. Meteorology provides the best and most vivid example. The weather, taken locally, never repeats itself in exactly the same pattern. Edward Lorenz showed in a computer simulation that, where weather is concerned, there is a sensitive dependence on initial conditions; that small changes in atmospheric conditions become magnified; that the results vary, are aperiodic and hence unpredictable. Yet such irregularities of weather patterns repeat themselves, and on a very large scale a regular pattern does set in. One could say that the weather is "locally unpredictable but globally stable" (Gleick, 1993).

In other words, globally the weather shows stability, in so far as the major seasons of the year and the geographical parameters determine the climate of the particular areas and the major weather patterns over the globe. This in no way ensures the accuracy of weather forecasting for a locality or even a region. In this sense, economics may be regarded as a close relative of meteorology. If we take the major economic parameters, such as the level of industrialization, the existing natural resources, the demographic state, and the political, social and economic structure of society, we find that these determine the kind of economic activity likely to prevail in different regions of the world. Yet the forecasting of the day-to-day fluctuations of stock prices or the duration of business depressions tends to enjoy a very low degree of success.

Returning to Gleick's colorful overview of the element of chaos in nature, it seems to us that in some respects he has overstated his case. The study of 'chaos' does not produce a new science, but it is true that it does probe more deeply into reality. Scientific work, through experiments, mathematical modelling, computer simulation and theorizing, does uncover the fantastic complexity of natural systems. This is highlighted by the many dimensions that reality has, and especially by the intertwining of stability, ordered patterning and regular change, on the one hand, with sudden outbursts, random behavior, turbulence and unpredictable bifurcation on the other. It appears, however, that in the face of this vast complexity there *is* in the scientific view also a more solid facet emanating from the limits, boundaries, categories and conditions which encapsulate the processes in nature, and which provide a framework for reality in its entirety.

Limits, Boundaries and Conditions

At this point the question that looms large is whether the limits and conditions are imposed, so to speak, by the very nature of reality, and uncovered slowly by scientific work, or whether they are what scientists *imagine* to be the limits and conditions, given the present understanding of reality. This question will remain unanswered forever. I believe this to be beyond human understanding. However, I shall try to illustrate what science views as the limits and conditions within which it attempts to explain what goes on in reality, in nature.

A boundary line that exercised the minds of thinkers and scientists from earliest times is the division between dead matter and living organisms. How rigid is this distinction? How does modern science treat this basic difference which to the layman is so obvious, the difference between the inanimate world and that which lives: rocks, water and air versus living plants and animals? There is no difficulty in demonstrating the difference between the living organism, even the simplest one-cell bacterium, and, say, a salt crystal. The latter is lifeless, the former is alive. Making the distinction in a different incisive fashion we may say with Richard Dawkins that "cells beget new cells, but dewdrops do not beget new dewdrops" (Dawkins, 1988). The essential difference then is reproduction. To quote a modern scientist on this topic: "One of the characteristic features of a living organism is its ability to reproduce itself. Living creatures reproduce, whereas inanimate objects do not. The reproduction of the organism depends, ultimately, on the reproduction of the individual cell. This is called cellular replication. The mechanism for cellular replication is identical in all living organisms" (Aviezer, 1990).

We shall not go into the subject of how reproduction takes place—how the genetic code works, what are the mechanisms through which replication is carried out, the organic chemical compounds involved, the structure of the cell, the possibility of mutations, the conditions necessary for the survival of cells, the reaction of a cell to a change in its surroundings, and cell death. All these aspects are well understood, and modern science has reached a very high level of knowledge concerning cell structure and the complex process of replication. However, it still has two major difficulties: (a) to explain the existence of borderline cases and some 'errors' that occur in biological processes; (b) to understand how living organisms have originated. One borderline case is a living creature that does not reproduce: the mule, a cross between two species that do reproduce, cannot continue the process of reproduction, it cannot propagate. Although basically the replication system is the same for all living organisms, the organization of the system varies between species, hence there is a "species barrier". The 'contents' of the genetic components also vary within species, hence individuals do differ from one another (Dawkins, 1988).

From the scientific point of view a more interesting borderline case, and one that is of immense importance for human beings, is that of the virus. It is questionable whether one may call the virus a simple cell: although it has some hereditary material, it cannot replicate on its own since it has not enough cellular machinery to do so. Hence we have a borderline case between the organic and the inorganic, and science has yet to understand its true essence and so discover how to deal with it, in order to prevent it from causing the dysfunctioning of organisms. Similarly, science is grappling with the most difficult problem of error rates in cell reproductions and spontaneous mutations.

The greatest scientific enigma is, however, how life originated on earth. Evolutionist scientists have made some attempts to show that living organisms could have developed from inorganic matter. A very well known figure, the Russian bio-chemist Aleksandr Oparin introduced his own idea in the 1920s that primordial organisms appeared out of existing *organic* compounds (*Encyclopaedia Britannica, Micropaedia*, 1992). These compounds, however, according to some scientists, could have originated from inorganic crystals. These scientists adduce two arguments in favor of their ideas: (a) using an electron microscope, one can observe how crystals grow in clays and mud; (b) on primeval earth electric discharges of lightning and ultraviolet radiation have produced organic molecules out of the inorganic crystals—something they claim has been demonstrated in experiments simulating such primeval conditions. Even if this could be achieved, another transition would be necessary, a transition from organic chemistry to biology. In 1953 Stanley L. Miller attempted to do just this in laboratory simulations of the beginnings of life on earth. However, as Julius Rebek remarks: "No one can say for sure what the atmospheric or terrestrial conditions were at the time or which molecule crossed the critical frontier" (Rebek, 1994). Apart from strong doubts as to the possibility of simulating conditions similar to those on early Earth, even evolutionist scientists do not claim a real knowledge of how living organisms originated.

In fact, in the last two decades, there has been a total reevaluation of life's origins and development. This has been due to the work of paleontologists who study the life of the geologic past through the analysis of plant and animal fossils. As G. Schroeder puts it: "There is now overwhelmingly strong evidence, both statistical and paleontological, that life could *not* have started on Earth by a series of random chemical reactions. Today's best mathematical estimates state that there simply was not enough time for random reactions to get life going so fast as the fossil record shows that it did" (Schroeder, 1990).

Richard Dawkins, who argues in favor of an evolutionary theory of life's origins, nevertheless emphasizes that there is a major problem: "The missing link...is still the origin of replication. The building blocks haven't come together to form a self-replicating chain" (Cairns-Smith, 1985; Aviezer, 1990; Dawkins, 1988). That is to say, even if inorganic matter can be transformed into organic substances, as such synthesizing in laboratories has been done since early 19th century, and even if synthetic self-replicating molecules can be produced in the laboratory, as Julius Rebek has claimed more recently (Rebek, 1994), it is still unknown what is the 'power' that organizes such substances into a reproductive living organism.

This unknown link is filled in by the philosophical doctrine of *vitalism,* which holds that there is a non-material 'force' in living organisms, represented in human beings by *the mind*, which is said to animate the bodily organism. But vitalism, just as biological mechanism, cannot explain the origin of the replicating organism (Broad, 1960). All we can do is lay down a ground rule or limiting scientific notion that there is a uniform genetic system which determines the continuity of life through self-replication. Another ground rule concerning living organisms is that they are open systems, meaning that they receive energy from outside, which by interconversion becomes useful for their functioning, and a basic necessity for their existence. The interconversion process of heat, light, chemical, electrical and mechanical energies, is that which makes biological organic life possible.

All living existence as well as all elements of which the reality of the universe is made up are dependent on certain conditions and physical limitations, which scientists aim to explain and define. One of the most spectacular physical limits of universal existence is *light*. The utmost importance of light, which is electromagnetic radiation detectable by the human eye, is appreciated both by laymen and scientists. If we imagine ourselves in a pitch-dark room with a sudden opening that lets in light, we would be wrong if we were to think that light entered our room *instantly*, and we would be partly wrong in thinking that it was now simply our brain and eyes that enabled us to see the objects outside or inside the room. In the first place, it took light a tiny fraction of time to enter the room; but since light travels at a fantastic speed it seems to us as an instantaneous phenomenon. Second, it was the transmission of information concerning the objects outside and inside the room, by means of light, to our eyes and brain, that made possible the 'seeing' of the objects.

The propagation of light was described a century ago by James Clerk Maxwell in terms of self-reinforcing electric and magnetic fields. Modern scientists regard light as electromagnetic energy which always moves, unlike other forms of energy stored, say in coal or oil, which have to be transported. Not too much imagination is needed to realize that beams of light from whatever source (the sun, the moon, or from artificially created electricity) shine onto an object, these being reflected from the object to our eyes, and thus the brain absorbing the information carried. Experiments have shown that optical instruments, such as spectroscopes, extract much more information than our visual system can extract from beams of light (*Encyclopaedia Britannica, Micropaedia*, 1992). This basic communicative function of light is stressed by many scientists. Let me illustrate this point by quoting J. Bronowski: "Light and other radiations are signals that spread out from an event like ripples through the universe, and there is no way in which news of the event can move outwards faster than they do. The light or the radio wave or the X-ray is the ultimate carrier of news or messages, and forms a basic network of information which links the material universe together. Even if the message we want to send is simply the time, we cannot get it from one place to another faster than the light or the radio wave that carries it" (Bronowski, 1973).

The great importance attached by scientists to light consists in the premise that the velocity of light is a universal constant, and a limiting boundary. Experiments have shown that the speed at which light travels is constant, irrespective of its source, and irrespective of whether the source or the receiver of the light are moving or stationary. This velocity—that is the speed at which it is transmitted—has been calculated to be 186,281 miles per second. Scientists have also laid down that its speed of transmission is the ultimate or top limiting velocity in the universe. It cannot be surpassed. This assertion is based on Albert Einstein's view of mass as simply concentrated energy. Therefore, matter and energy are basically the same; they are interchangeable. As one scientist put it: "If matter sheds its mass and travels with the speed of light we call it radiation or energy" (Barnett, 1968). Or as Hawking puts it: "Only light, or other waves that have no intrinsic mass, can move at the speed of light" (Hawking, 1988). Consequently, in the case of any object, the ever increasing energy needed for it to move nearer to the speed of light will add to its mass, yet this increased mass will need ever more energy to move it at such a tremendous speed; that is to say, it would need that which is impossible—an infinite amount of energy.

Hence all objects which have mass can move at slower speeds relative to that of light. The mystery is still there. What does Hawking mean by "no intrinsic mass"? That the phenomenon of light or other radiation transmission is possible at this tremendous top velocity because of its dual nature as both wave and particle? Can we, therefore, say that its transmission is in the form of waves, which has no intrinsic mass, yet in its essence or at its source and at the stage of its absorption light, too, is composed of particles and it behaves as such?

In fact the explanation of what happens in the interaction between light and matter was given in 1905 by Albert Einstein (1879–1955), who held that although light has a wavelike characteristic, it can be regarded alternatively as composed of discrete particles which are called *photons* and are equivalent to energy quanta. The interaction between light and matter produces the *photoelectric effect*, a phenomenon in which particles are released from a material which absorbs radiant energy. The way was prepared for Einstein's theory by J.J. Thomson, who earlier suggested a discontinuous theory of light, and by Max Planck, who held that heat radiation is emitted and absorbed in distinct units, or quanta. These ideas were generally accepted after the American physicist Arthur H. Compton showed in an experiment in 1923 that light "displays both wave and particle properties simultaneously" (Trivers, 1985; *Encyclopaedia Britannica*, 1992).

Light's existence as waves belongs, therefore, to that exceptional world of wave transmissions, of light or radiation, at the super finite velocity measured by science. There have been, however, some astronomers' reports of having observed a blob of matter in our galaxy doing the impossible—moving at the phenomenal speeds of nearly that of light, at about 92 percent of the speed of light. Various observations of even faster-than-light movement outside the Milky Way galaxy have been termed "the faster-than-light illusion"[3] It is obvious that further research and a better understanding of the illusion is necessary. The knowledge so far accumulated is insufficient to challenge the universal physical boundary we know as the speed of light.

This spectacular physical limit that light imposes on universal existence is just one of the underlying mysteries that has left scientists baffled, however sophisticated their explorations and explanations of what goes on in reality might be. Another universal phenomenon, the force of attraction that affects all matter, gravitation, is still not fully understood by scientists, despite all the distinguished work done from Newton to Einstein and beyond. They are now attempting to produce a 'grand unified theory' to explain all the fundamental forces in nature (Munitz, 1986).[4] And again, as our brief review above has pointed out, scientists have still not been able to tell us how organic life originated and what is the 'power' that organizes its reproduction.

In trying to make sense of reality we have been taken along some tortuous scientific paths which do not lead us to rock-solid answers to some of the more basic questions. If we try to deal with the most basic notions concerning the universe, such as space and time, we shall slide deep into metaphysics—into sheer philosophical speculation. As Leibniz said nearly three hundred years ago "universals have no reality but are mere names" (*Encyclopaedia Britannica, Micropaedia*, 1992). Thus, space and time have nothing substantive about them; they are just concepts of the human mind, trying to express within some framework our very finite existence and the existence of the infinite universe. On this view, space is a concept that encapsulates all the substances and existences that ever existed and that will ever exist; while the concept of time expresses all the events that ever took place and will ever take place. Space and time in this sense are not something real or actual, and the universe in this sense is infinite and timeless.

Yet there *is* a reality that, as Karl Popper said, "kicks back at you" (Popper, 1972). It is this 'real stuff' that the scientist deals with. And if there is need for a spectacular demonstration of scientific reality, we have only to refer to an event which occurred on July 16, 1945 near the small lumber town of Almagordo in Southern New Mexico, USA, situated at the western edge of the Sacramento Mountains. On that date, at the "Trinity Site" of the Holloman Air Force Base, the first atomic bomb was exploded. Thus, "man for the first time transmuted a substantial quantity of matter into the light, heat, sound, and motion which we call energy" (Barnett, 1968).

Although the fundamental mysteries remained and perhaps were added to by this at-once creative and destructive giant step in the scientific revolution of the modern age, one could put forward the practical aspect of science, which may be summed up as knowledge of the way things are caused. It is to this aspect of causation that I turn my attention in the next chapter.

NOTES

[1] Readers who are interested in astrophysics can turn, for instance, to the writings of Stephen Hawking, in order to follow the theories relating particularly to the question of entropy in connection with "black holes", which are regions in the universe where the gravitational force is so strong that nothing, not even light, can escape from them (Hawking, 1988).

[2] DNA (short for deoxyribonucleic acid) molecules are made up of sugar-phosphate chains linked through nitrogenous bases by hydrogen bonding. Such molecules of a particular organism are capable of self-replication. Using the language of the genetic code the DNA determines all the inherited characteristics of the organism—it is the blueprint handed from generation to generation.

[3] One such report was communicated by the Associated Press from New York (2.9.1994). A similar sighting, claimed by some scientists, of "a faster than the speed of light" phenomenon, and speculation as to whether interstellar travel may become possible, was reported on the 6 o'clock London BBC 1 TV news on the 7th of September, 1995. It has been suggested that a small probability exists for particles in some conditions to travel faster than light; physicists have attempted to measure speeds of such particles in experiments (*The Times*, 3.4.95). A report by an astronomer submitted to the Royal Astronomical Society in London suggests the

possibility of faster-than-light space travel. However, Sir Martin Rees, the Astronomer Royal and Professor of Astronomy at Cambridge University, was more cautious, saying that "the proofs were purely theoretical" (*Sunday Times*, 13.8.1995).

[4] For a detailed discussion of attempts to discover a fundamental law or a 'theory of everything', see: Gerard 't Hooft, 1994. On the quest for a 'final theory' see Steven Weinberg, 1993. Concerning the universal phenomenon of gravitation, it may be noted that in recent years some researchers have tried to find a device to combat gravity on Earth, which could spark a technological revolution, while other scientists are planning a gravity probe in space which will test Einstein's theory of general relativity (*The Sunday Telegraph*, 1.9.1996, p. 3; *New Scientist*, 31.8.1996, pp. 28–31).

CHAPTER THREE

SEARCHING FOR CAUSES

Science provides knowledge which includes both a description of our lives and the world that surrounds us, as well as their dynamic aspects: the way life and the world change, and the way the various elements in our lives and in our world affect each other. In this sense science's aim is to provide a knowledge of causes.

Here we must indulge in some semantics. A number of specifically defined concepts can be used to relate to different aspects of causes. Thus, I could ask: (a) *what* caused X; (b) *how* has X come about—that is, the process that was involved in the causation; and (c) *why* has X occurred at all? Let me illustrate with the example of the sonic boom. Taking the first question, I could answer simply that when a supersonic aircraft breaks the sound barrier (that is, the speed of sound at the particular height at which it flies), it produces a thunderous sound on the ground. As to the second question, my answer will be somewhat more involved: an aircraft that flies below the sonic speed produces sound waves which outspeed their source and spread out ahead of it; but once the aircraft reaches sonic speed, the waves that it creates are unable to get out of its way, which results in tremendous friction, heat and turbulence in the air, and in the boom we hear on the ground. The third question is the most difficult however. We are asking now *why* the sonic boom has occurred, not in the sense of why on a particular occasion it happened. The answer to this could be simply that a pilot named Bill had taken up an aircraft and had flown it at the height of 12,000 meters at over 1,050 kilometers per hour—that is, at over Mach 1—thus breaking the sound barrier. Our 'why' question concerns something much more basic and general: why does sound travel at the speed at which it does, thus producing a "sound barrier" to a supersonic aircraft which by overtaking it creates the sonic boom?

The semantic distinction we have just used in asking questions concerning causation may be summarized as follows. The *what* question emphasizes the time/sequence in causal link, the effect always coming after the cause. The *how* question deals with the conditions which enable such effects to occur, and the ways in which the elements or events interplay and thus bring about the effects observed. Here science merges with technology. The *why* question touches on the philosophical issue of first causes or the ultimate answer, beyond which there is no possibility of further inquiry or analysis.

Most scientists hold the view that science never asks 'why' questions in the sense of ultimate causes, and that questions which cannot be tested in reality are not scientific questions, interesting as they may be. On this view science is not concerned with fundamental questions regarding what is life or the beginning and the end of the universe. Rather, detailed questioning of the world around us, given to observation and testing, gave rise to more penetrating and generalized explanations (Weisskopf, 1981). In this sense science restrains itself, it sets its own limits. However, I follow the broader view, in this chapter and in other parts of the book, that more fundamental questions in science are unavoidable. It is true though, that general conceptualizations, such as the unity of time and space, and questions about the origins of life and the universe, bring science to its limits. Nevertheless, our attention will be directed to the various facets of scientific explanation emanating from attempts to deal with the type of questions we posited, including the more fundamental ones.

Causes and Facts

It is a truism that the effects we observe are facts. What we are after, though, is to find the facts that caused the effects. In other words, we wish to link various facts in a causal manner. For example, suppose we observed that a china-plate has broken into bits as it landed on a hard surface. We have just observed a fact. The question is whether we also observed some other fact, such as seeing someone in bad temper throwing the plate, or feeling an earthquake and seeing the plate slide off a tilted shelf. In either case we have no problem in stating that our causal explanation is based on facts which occurred according to a temporal sequence.

But let us go just a little beyond these simple observations. We shall find that our causal explanation involves Newton's law of gravity, if we choose to apply it, according to which the gravitational force of the earth (and of other planets) exerts an attraction on objects sufficiently close to it. But this gravitational force is not something directly observable; it is, therefore, a concept or theory which enables us to explain more fully what caused the plate to break. Furthermore, we could also propose a counter-factual statement as to what would happen in other circumstances, say, in a spacecraft travelling in outer space. There, in the absence of gravitational pull, a plate which is thrown does not necessarily land and break. Our initial causal explanation is, therefore, not applicable universally. One more point here: not only are theories, based partly on non-observables, necessary for causal explanation, but such theories change in science. Thus, the Newtonian gravitational force idea has been overtaken by the Einsteinian gravitational field concept—that the presence of such fields around large masses result in the curvature or warping of space near the masses, which affects objects including planets or light rays. "A simple analogy is the way in which a heavy ball sitting on a rubber sheet distorts it, making an indentation so that any small objects in the vicinity tend to roll towards it" (Iannotta, 1996).

Now, it seems reasonable to assert that neither 'forces' nor 'fields' can be observed in themselves, but it is true that scientific evidence can be adduced to show that their initially assumed effects actually take place. This is what happened in the famous case of Einstein's general theory of relativity. The latter holds that gravitation is not a force, as Newton held, but that the presence of matter in space causes space to 'curve' in such a manner that the gravitational field is set up. Thus gravitation becomes a property of space itself. This theory predicted that light rays would be bent by the gravitational field of the sun, as against the straight-line prediction based on Newtonian physics. Eclipses of the sun in 1919 and in 1922 provided the opportunities for Arthur Eddington and William Campbell to photograph the paths of light rays of certain stars when in the vicinity of the sun. The results showed that, indeed, the rays were bent and thus the new theory of this phenomenal scientist was verified by observation (Hawking, 1988; Lieberson, 1992; *The Penguin Dictionary of Science*, 1993).

Despite this spectacular scientific evidence which, after more than two centuries, basically modified Newtonian physics, it is still true to say that 'the cause' of the bending of the rays—the gravitational field—cannot be observed. To emphasize this point further: the sun can be observed and so can the earth moving in its elliptical orbit around the sun, as well as the bending of light rays in the vicinity of the sun, but the special *property* of space implied by the gravitational field idea can certainly not be observed.

Similarly, the more recent idea that the gravitational force between the sun and the earth is ascribed to the exchange of *gravitons* between the particles that make up these two bodies, does not solve the problem of the unobservable. Quite apart from the fact that gravitons, by the admission of modern physicists, "have never yet been observed", and thus remain hypothetical particles or quanta of gravitational energy (Hawking, 1988; *The Penguin Dictionary of Science*, 1993), the essential gravitational quality of such gravitons, if they do exist, would remain unobserved. It is in that quality, one imagines, wherein lies the true *cause* of gravitational attraction, so that neither on the surface nor deeper down can the assumed cause be observed. Science is replete with such examples. In fact "within the standard model of particle physics, some types of interaction are conceivable but in practice never seen" (Cline, 1994).

Yet, we could say that our perception of causation is not really undermined by such unobservability, so long as we modestly agree that scientific descriptions of causal processes or findings, as David Hume, the famous eighteenth century Scottish philosopher, held, are simply observed regularities. Scientists, however, are not content to observe regularities; they wish to provide causal explanations. This is also true of the layman who tends to give some explanation for ordinary occurrences in his life, such as: "I fell down because a paving stone was out of place". The scientist will try to provide a causal explanation for his observations of natural phenomena, often aided by powerful instruments, or for the results occurring in his artificial experiments. Thus, a scientist may observe with regularity that magnetite, an iron oxide mineral, has the property of picking up pieces of iron and will explain this by the ferromagnetic theory. What he is saying is that the iron oxide's magnetic force caused the other metal to be attracted. As in the case of the gravitational "force" or "field" mentioned above, or the "loss of balance" induced by the faulty paving stone, so in the case of "magnetism", we have reasonable explanations of events or processes, which are made possible by our connecting different elements of reality in a causal manner. Therefore, at this point it will be useful to say in what way and in what sense we make such causal connections.

Causation: Ideas and Definitions

There is a huge literature on this topic written mostly by philosophers of science. As background, let us quote one writer of the nineteenth century. Thus William Whewell: "every event has *some* cause [which] experience cannot prove any more than can disprove;...we do in fact apply the Idea of Cause in a more extensive manner than could be justified if it were derived from experience only;...by the use of our powers of observation, we are aware of a succession of appearances and events. But none of our senses or powers of external observation can detect in these appearances the power or quality which we call Cause. Cause is that which connects one event with another, but no sense or perception discloses to us, or can disclose, any connexion among the events which we observe. We see that one occurrence follows another, but we can never see anything which shows that one occurrence *must* follow another;...there is no such discoverable necessity". By stressing that by cause we mean not only "a constant unvarying succession" which we observe, but also the implied non-observables which he terms "a power, an efficacy, which has a real operation", he arrives at the axiom that: "Nothing can take place without a Cause" (Kockelmans, 1968).

As A.J. Ayer says, the above proposition that every event has a cause expresses "the principle of determinism" (Ayer, 1962). Thus, one can define causality deterministically by saying that the fulfillment of certain *necessary* and *sufficient conditions* will always be followed by the occurrence of an event; say, an electric discharge coming in contact with inflammable material in a dry space containing sufficient oxygen will cause a fire. It is not clear in this kind of definition how we distinguish the causal factor (say, electric discharge) or factors (say, spark plus oxygen) from the rest of the necessary conditions, which together constitute the sufficient conditions for the fire to be caused. Be that as it may, what the above kind of definition stresses is the apparent constant conjunction between cause/condition and effect, that is the unvarying succession involved (Hospers, 1956).

According to Ayer, the distinction between cause and condition is arbitrary but not entirely capricious. There are relatively stable conditions, in addition to those that change—it is the latter that are usually singled out as causes: "Thus in the case of a forest fire, it is the spark that ignites the fire and the wind that fans it, that are said to be the causes of the conflagration, rather than the state of the climate or the composition of the wood" (Ayer, 1972).

Another more rigorous formula for causation is: 'If C happens, then (and only then) B is always produced by it'. This definition does not assume that constant conjunction and constant succession show causation, since both could be accidental. In this formulation "the gist of causation is uniform *production*"; this idea of production is further sharpened by Mario Bunge, whose view is that the causal bond is apparent only when A *produces* B (Bunge, 1963).

Now, do I see and taste the causal bond when I mix milk into my coffee and drink the brownish liquid? In a sense the production of the milky coffee was apparent to me, but I may need a professional chemist and his instruments to tell me something about the causal bond, in terms of the mixing of different molecules. By using very fast photography or reflective spectroscopic methods, he could observe that the mixing of milk into the coffee liquid produces a diffusion of the milk molecules, this in turn bringing about a transient bonding, and as a result producing a change in the spectrum, that is, changing the liquid from a deep brown to a lighter brown, and also changing the taste, both of which I pick up with my senses.

There is no doubt that up to a point one can observe the causal process taking place, especially with the help of science; and the layman can sense the effects of the process. Yet, there is a point beyond which present scientific knowledge falls short of a coherent description of how causes are related to effects, such as in the case of events and processes occurring on the sub-atomic scale. Whilst molecules and atoms and their interactions can be observed, sub-atomic particles pose a limit on observation. Limitations on observation of causal processes are even more severe in the social sciences, as we shall see below in Chapter 6. In the world that the social sciences deal with, events often have such a complex causal background that the actual process or processes leading up to such events are well-nigh impossible to disentangle, and certainly impossible to observe.

The idea of a causal process leads us logically to the need to acquire scientific knowledge, by means of observation, the application of hypotheses, experimental testing where possible, and theory-aided scientific analyses, in order to understand the causal process. This means that we must look not only at immediate production, such as my pumping more gasoline into my car engine and thus producing greater speed for my car, but also at the way the entire process of the internal combustion engine works so as to result in moving the car and in increasing its speed, something that one could call propagation. Thus, energy emanating from the ignited fuel is converted by means of an enclosed cylinder behind a piston into mechanical energy which moves the car. It is the transmission of the energy that can be identified as causal propagation. Such transmission/propagation occurs in thousands of cases, one example being the electromagnetic waves that transmit sounds from a radio station, which are then received on our radio sets at home. A process of transmission/propagation also occurs where the memory of a human mind of past events affects causally his behavior or attitude at the present moment (Salmon, 1988).

Obviously, causal processes could be shortish or of very long duration and persistence. Bertrand Russell, the famous twentieth-century British philosopher, paid attention in particular to the persistence of self-determined causal processes, which he called 'causal lines', these being defined as "temporal series of events so related that, given some of them, something can be inferred about the others whatever may be happening elsewhere" (Russell, 1948). The best example is probably a person, whose 'line' of development from childhood to adulthood and to old age is a persistent causal process of gradual change with a certain constancy of biological structure and mental quality. Thus, given the baby's sex we can infer with a great deal of certainty his subsequent biological development.

Another interesting idea on this topic is that of *causal chains*. In history and sociology, for instance, we often talk of a causal chain of events, whilst in the natural sciences the idea of chain reactions is very common. Probably the best known example of chain reaction is that which occurs when splitting the atom. This is known as nuclear fission and is achieved by bombarding the atom with atomic particles, such as neutrons. In the case of the uranium atom such fission will release more neutrons, which in turn will split more atoms thus releasing an ever increasing number of neutrons. In short, a tremendous chain reaction occurs as a result of which a vast amount of energy, heat and powerful radiation is emitted.

A state often experienced by human beings, such as thirst, is due to a physiological causal chain. Briefly, a lack of water in the body affects the salivary glands, which are unable to secrete, thus resulting in the mouth and pharynx becoming dry, due to which the sensation of thirst arises. The causal chain clearly implied by this statement is: deficiency of water in body tissues leads to a reduction in salivation, which leads to the dryness of the tongue and palate and thus to the stimulation of the nervous system, giving rise to the sensation of thirst (Simon, 1977).

In proposing the idea of causal chains we imagine reality to be composed of discrete events which are linked. A popular imagined historical account of such a chain would be: "For want of a nail a shoe was lost; for want of a shoe a horse was lost; for want of a horse a rider was lost; for want of a rider a battalion was lost; for want of a battalion a battle was lost; for want of a victory a kingdom was lost—all for want of a nail" (quoted by Hanson, 1961). One can imagine, however, not just one causal chain ultimately bringing about some occurrence, but rather the meeting or intersection of two or several causal chains that produce the event, occurrence, etc. A simple example would be where we can designate event one, the parents sending their child X to school; followed by event two, the teacher telling the child to sit next to another pupil Y, who has a throat infection; followed by event three, the child X contracting the infection. But inquiring further we may find that pupil Y was infected at home by his sister, who in turn was infected by a colleague at work. Thus the meeting of two causal chains produced the occurrence in question, the infection contracted by child X.

An historical example of the intersection of causal chains could be given when the causation of the Holocaust is explored. Here one can imagine one historical chain (A) linking the conquests of the Roman empire with the exile and dispersion of Jews in Europe two thousand years ago, followed by their reluctance to integrate into the newly created Christian culture, which left them segregated for many centuries, and which in turn produced sociological and psychological processes resulting in antisemitism. The other historical chain (B) could start, arbitrarily, with the economic and political struggles in Europe leading to World War I, and the economic depression which paved the way to the utlra-nationalist Nazi ideology and the Hitlerite dictatorship, the latter exploiting the existing fertile ground of antisemitism and implementing the so-called 'final solution', the colossal catastrophe which was the Holocaust. Had chain B not been in process, or had it not followed exactly the path it did, the Holocaust may have been avoided. Similarly, had one of the chains in the case of the infection not taken the path it took, child X may not have contracted the infection on that occasion.

The major question is whether scientific research can usefully apply the idea of causal chains. Such causal chains have a number of characteristics which seem problematic for science: (a) causal chains seem to link *accidental occurrences*, for instance the ball I kicked seems to have been stopped accidentally by a big stone on the field; or to take the idea of causation resulting from the intersection of two or more causal chains, we could again invoke *chance* as a factor, unless we are pure determinists, believing that since the intersection of events, linked through causal chains, has produced an occurrence, this in itself is the 'proof' that the occurrence *had* to take place; (b) each one of the events/things linked in a chain is itself part of a larger structure of reality; thus a deeper and fuller scientific analysis, it could be argued, should be instituted for each event in its own context: for instance, the proclivity of child X to contract infections must be studied in the context of his entire physiological make-up, quite apart from the question of how he came into contact with the source of infection; or antisemitism in Germany requires the examination of the entire cultural development of German society, as well as the even wider topic of Jewish-Gentile relations over many centuries; (c) causal chains can be traced backwards so far as to make them nothing but imaginary, leading to scientific speculation rather than realistic scientific study: thus the ancestry of any human being today is, in this imaginary sense, traceable to the first human being on earth, and even to an earlier phase of the development of life on earth; and any tiniest particle in the universe has its origin in the very first moment and the very source of the creation of the universe, and, logically, the antecedents of that creation *ad infinitum*.

Despite these caveats, scientists do design their experiments in a chain-like fashion. Of course, they will select out portions from longer sequences imaginable; those portions which focus on their topic of interest. Furthermore, the extrapolation is made out of a vast number of possible sequences that could have been selected for studying, including the sequence of the occurrence of interest. The laboratory experiment will test the particular sequence selected by holding constant all other sequences. The scientist interested in the effect chemical A has on chemical B at a certain temperature, will hold the temperature constant as specified by him, since the natural sequence of the temperature such as a warming process effected by sunlight may alter the behavior of the chemicals and thus destroy the sought experiment.

A scientific experiment poses a paradox since it selects out special sequences, and thus in a sense destroys the natural links that objects and events have in reality, by decomposing nature into events and particles. Alfred North Whitehead drew attention to this paradox and solved the problem by declaring that all the objects and events we perceive whether in the 'natural' world or in the 'scientific' world are abstractions. In both the realm of *continuous* links of objects and events in reality and in the *discrete* collection of components and the scientific attempt to understand the causal connections of these components, in the last resort all we do, through sense-data, is to relate to and connect various objects and events in particular situations (Palter, 1960; Whitehead, 1920).

On the whole it would be reasonable to say that the longer the causal chains proposed, the more imaginative and the less scientifically supported they are. An excellent example is Charles Darwin's theory of evolution, according to which "all kinds of living things (including human beings) developed from very few simple forms through natural selection", which worked as a "process of gradual change in the characteristics of organisms over many generations" (*A Dictionary of Philosophy*, 1979; *The Penguin Dictionary of Science*, 1993). Here we have a causal chain with a phenomenal number of links in it, for which no serious scientist today would claim the possibility of finding clear evidence (Gould, 1984). In fact the fossil evidence shows the contrary: instead of gradual change, scientists say, the evidence suggests the sudden catastrophic extinction of species (Aviezer, 1990). But some scientists hold that Darwin's ideas may still be useful and serve as a general proposition or paradigm in modern science, which can take as a sample a shorter stretch of the extraordinarily long Darwinian chain. Focusing on what I called a shorter stretch would also involve very long periods. The main point, however, is that attention is given to a particular aspect of evolutionary change, where scientific evidence is easier to adduce. Thus the scientist can focus on the particular protein of hemoglobin found in abundance in the red corpuscles of the blood, and on the changes that took place in this substance as seen in the different species, and which then became more or less preserved in them. If one compares the genetic sequences of hemoglobin in different species, one finds that the differences between them can be explained by an evolutionary process, which apparently arises from mutations that occurred over generations. In this fashion one can construct a phylogenetic tree for the different species, showing a classification of species and the relationship based on a closeness of evolutionary descent (Cantor and Schimmel, 1980). On the whole, however, the idea of evolution has lost a great deal of support in science. It is alleged, for instance, that James D. Watson the co-discoverer of DNA "openly expressed contempt for evolutionary biology" (Horgan, 1994; Milton, 1992).

While my main aim is to convey to the reader the general idea of causality, I do not wish, by default, to give him the impression that scientific thought runs exclusively on the idea of linear chain-like sequences of causation. It is necessary, therefore to mention, even if only briefly, other models and ideas through which causation is analyzed in scientific work. Thus, a model of *multiple causation* may be used where the effect is produced by several causes, independently or in combination. A simple example is where the flow of a certain volume of water results from a single big source, or from several lesser sources. There are also cases where a single cause gives rise to *multiple effects*. Again to exemplify, an electric bulb has the effect of lighting up the environment and also of warming it. An interesting case is that of *mutual causation*, where some factor or variable "is not just passively dependent on its causes, but, in turn, it influences one or more of its determinants" (Heise, 1975). Where such non-linear reciprocal causation exists, models in the form of *feedback loops* are used. For instance, X = additional money invested in industry, provided that the additional necessary production factors exist, produces Y = more employment, which in turn puts more money into the economy, part of which is saved and some re-invested in industry, thus raising the volume of X through the feed-back effect.

Causal Inference

We must now ask the basic question whether one can use the concept of causality in cases where the causal structure cannot be fully demonstrated. It is not always possible to show the causal connection between cause and effect. Such a connection is obvious in the case of a system involving a clear process, say my turning the engine key igniting the fuel and moving the car. But often we can only *infer* the connection, as in the case of a group of picnickers who after a day in the country become violently ill that evening, where we guess that the common meal they ate must have caused their illness (Salmon, 1988). Of course, it might be possible to examine the patients and the food they consumed, which might reveal the poisonous stuff that caused the picnickers to become ill.

In some cases, however, even this is not possible; there is no way of establishing such direct links. All we have is covariation between types of population and types of behavior. Statistical data may show us that in the older age-groups a higher percentage of people vote for right-wing parties, whilst in the younger age-groups a higher percentage vote for the left-wing parties. From such statistics we can infer that age, or more correctly experiences due to age, has an effect on political behavior. The inference will lead, however, only to probabilistic explanation of how age influences political behavior, since we shall find quite a number of people who do not fit the pattern. In all cases where causal explanations are sought through statistical correlations, whether in the case of the use of antibiotics which work for the vast majority but *not all* the patients, or the drop in the number of marriages in time of war, economic depression periods, or other crises, the reason for the particular distribution of outcomes is due to chance or choice playing a role in the events. Thus, by chance a number of patients have a physiological make-up which is not suitable for the particular medicine used. Again, by choice various couples decide to get married despite the war situation. However, such lack of certainty does not destroy the causal inferences drawn on the basis of statistical data. But it does mean that alternative outcomes are feasible, and that regularities we uncover such as that in most cases infections can be cured with antibiotics will be probabilistic.

Another uncertainty in causal links inferred from statistical data is the possibility of spurious correlation. Students in first year statistics courses soon come upon the slogan that "correlation is no proof of causation". Herbert Simon describes the way we ordinarily make causal inferences from data which show correlation between two variables. Let us suppose that we see that a high percentage of female employees is positively correlated with a rise in the number of absences from work per week. Can we infer that the mere fact of being married causes high absenteeism among women employees? Scientists who seek to establish real causal effects would not accept such a simple assertion and may conjecture that a third confounding variable is at work. Indeed, if we hold housework constant the correlation drops to zero. In other words, housework is the intervening or mediating variable: that is, marriage results in a higher average amount of housework for the female partner and this in turn results in more absenteeism among women workers (Simon, 1977; Dunbar, 1995).

Even more daunting is the task of inferring the direction of the cause-effect relationship from correlational data. Let us take an example from the demographic field. At the end of the last century a high positive correlation was observed between low-income families and their birth rates. Which way did causation go? Was poverty an effect of the large size of families—was a large family unable to provide sufficient material sustenance for itself; or were the poorer classes either unable to obtain the fairly expensive birth-control methods available at the time, and/or were they unaware of the possibilities of birth control because of their low level of education? And is this why they continued to produce large families? We seem to have here a classic example of mutual causation and reinforcement.

Since often correlational data do not point clearly to causation, we may be able to point only to indirect links or very remote connections. For instance, common sense would suggest that a statistical correlation between the growing quantity of sugar consumption and the rising rate of divorce is no proof of a direct causal link. Both may be said to be the effects of a modern life style, but certainly one cannot infer from this that those who consume more sugar are more likely to divorce. By the same token we may find a correlation between the increasing number of office-window cleaners in large cities, and the increasing amount of sugar consumed in such cities, and similar correlations may appear between cholesterol levels, divorce rates, urban crime, etc., yet no social scientist would make any causal inferences on the basis of such statistics. In a later chapter, devoted to the social sciences, we shall consider how social scientists select those correlations which are causally meaningful.

In the natural sciences, where it is easier to demonstrate causal links, the entry of probability into causal explanations and predictions has certainly weakened the principle of causality. From this angle it would seem that there are no iron-clad immutable laws of causality. Pierre Simon Laplace, the French mathematician and philosopher of two centuries ago, held however, that it is only our ignorance, our lack of knowledge, that makes it appear as if things happen by chance or at random. He claimed that "were he but supplied with an account of the state of the universe at any one moment, plus a list of all the causal laws, he could predict and retrodict every other moment of the world's history" (Hanson, 1961). In other words, according to Laplace all that exists and happens is at base due to existing causal laws. Such a view is strongly contradicted by those who adduce evidence from modern physics as to "the existence of *random* events represented by such irreducibly stochastic processes as radioactive decay". It appears that, according to currently accepted scientific theories, the decay of an unstable atom takes place without the operation of specific 'causes'. The decay may be regarded as a spontaneous, dare we say *uncaused*, occurrence. We are talking here about the probability of the decay, that is the transformation of radioactive substance due to a spontaneous disintegration process in which the nucleus of an atom emits particles, with an exponential decrease of the activity with time. Thus, stochasticity, the element of probability, has brought about, according to Nicholas Rescher, the downfall of causality (Rescher, 1970).

It seems to me that both views are extreme: on the one hand we do not have to accept unreservedly the classical determinist definition of causality, according to which events are absolutely necessitated in the manner of "If A then always B'; on the other hand neither do we have to accept the view that probabilistic explanation is not causal. An indeterministic situation can apply to many areas of inquiry in which A itself may be variable in its effect, so that only the *probability* of it producing B is observable in the way of statistical regularities. Thus, a certain quantity of rain over a particular type of terrain may or may not produce landslides; certain medicines may cure only a percentage of a population; and some traffic laws may reduce accidents to some extent. Such effects are brought about by certain causes and we do have in these cases causal explanations on a probabilistic level.

Determinists would argue that with more exact knowledge of the different types of terrain, of the precise physiological and psychological make-up of people, we could explain why the effects differ despite the occurrence of apparently the same causes. This kind of argument leads in the direction of a belief (a) that science could possibly achieve perfect knowledge; (b) that the world is perfectly ordered and in the last resort it is based on immutable laws. In the foregoing we have raised doubts concerning such beliefs, and adduced some evidence of disordered relationships between elements in our life-world and that changing patterns, probability and chaos, are integral parts of reality.

On the whole, a balanced view is one that regards causal laws as a global factor, and causation as a local one and given to uncertainty and chance. Thus, a specific conception is very much a matter of chance, not only in terms of the possibilities of this occurring between two partners on any one occasion of intercourse, but also insofar as the very relationship between the two having taken place, rather than with other partners. But it is also true that without the occurrence of a particular intercourse between that specific couple, the probability of the particular pregnancy that *has* taken place would have been zero. Consequently we can see that a causal law which explains the occurrence is there globally, but that an actual occurrence at a particular moment in the circumstances in which it takes place is given to chance.[1] The actual specific occurrence is a local event and we often wonder what would have occurred if things had happened differently at a given moment in time (Bigelow and Pargetter, 1990).

In addition to the various doubts I raised in the context of what we may call our 'causal world', I now turn to some even more baffling issues concerning causes.

Human Will, Final Causes and Beyond

There is no doubt that human will must be regarded as a cause of a vast number of objects that exist in the world and of events we witness. At the same time there are many changes that objects produced by human will undergo, and many stages in events originated by human will where human will no longer seems to play a role, except as a causal source. And there is certainly an abundance of things and events which are in no way related to human will.

To elaborate on those objects and events which were or are caused by human will, let us imagine that our planet was never inhabited by human beings. We can immediately visualize a totally different world: to start with, the landscape would be quite different, with ancient forests in great abundance, the earth not tampered with by human agricultural pursuits, rivers flowing in their natural river beds, not artificially diverted anywhere and their banks unaltered by human labor: no canals, no bridges and no river or ocean ports. The crust of the earth would have no tunnels bored through mountains, no coal mines, no oil wells or oil rigs, and no quarries. The earth's surface would not be cluttered with vast amounts of steel and concrete; the atmosphere would not be polluted by fumes from factories and cars; and there would be no industrial and nuclear waste. There would be no means of transport, no buildings or monuments and practically all the contents of the buildings would not exist. There would be no artificial light, whether emanating from fire or electricity. There would be no new chemical compounds, no plastic, no medicines, no machines, no implements; no communication, no computers, no kites or aircraft in the sky, no satellites in space, and no man would ever have set foot on the moon. There would be no human reproduction, no families, no social institutions, no music, no literature and no science. And obviously there would be no human language, religion, philosophy or art. The world would have no politics, no war and no crime as understood by human beings. The list is inexhaustible, for it expresses the vast amount of humanly produced artifacts and humanly caused events and ideas.

It is a moot point whether elsewhere in the universe creatures with something akin to intelligence exist, who have produced artificial objects and events. All I am trying to press home is that human will causes things to happen: it is an originator. It is evident that when I kick a ball across a field it is due to my conscious will, so that the motion of the ball was caused by human will. The ball might have been deflected by an obstacle and have taken a direction I did not intend. At this secondary stage my will has become diminished in its importance as a cause, the cause for the direction being the obstacle, and my will is no longer able to enforce the stages which follow. Human will may thus be seen as a first or original cause, but even this may not be a true account, since my present will may be a result of memories I have, my genetic inheritance, other previous events I experienced and circumstances which shaped my individual physique and psyche, and so on (Pearson, 1937). In other words, it will not be of much use in science to search for first causes in human will. Scientists must regard conscious will as one among the many causes that make our dynamic world tick. Very often conscious human will combines with other causal factors, whether natural or artificially created, in bringing about processes and in producing things.

But do these processes and things have, somewhere deep down and way beyond our present knowledge and imagination, some 'first' or 'final' causes? From a philosophical or conceptual standpoint it makes sense to ask such a question, to think of first or final causes. But does it make sense to do so from a scientific standpoint? Is it sensible or is it futile to ask 'why does a virus behave like a virus?', or 'why do sub-atomic particles react to each other the way they do?'. The pragmatic answer is that science must ask such questions if it wishes to dig deeper into reality and arrive at more knowledge and a better understanding of that reality. But it must ask such questions not in the metaphysical sense, but rather in the sense of what are the causes and how do the processes work, in order to effect such behavior. Science does this by proposing hypothetical explanations, to which it tries to adduce evidence from the objective world, and by applying this knowledge to reality. Such explanations are, as a rule, causal.

Such an aim for science is not to be seen as a commitment to pure empiricism and positivism, a conception that all knowledge of the objective world is based only on experience, and that no attempt should be made to go beyond observation. Science would stagnate if it were to adopt such a stance, for it is based on mind-dependent descriptions, concepts, hypotheses and theories. The very fact that scientists often postulate various hypothetical possibilities in explaining phenomena attests to a realistic approach. It admits to the existence of some objective world out there, which strictly speaking is not dependent on the human mind or on scientific knowledge, yet it uses human reason, intuition and imagination in the form of scientific theories and experimentation, to grapple with the mysteries of the universe.

Herman Cohen, the 19th century German philosopher, pointed more clearly to the interdependence between the objective world and human reason and logic. He regarded the dichotomy between 'facts' and 'theories' as artificial and unnecessary. After all, sense-data become spontaneously structured by human reasoning; our scientific facts are theory laden; and our starting point is "not the notion of the absolute, but the actual scientific achievements as a real, concrete dynamical context embodied in concrete problems and in their solution" (Funkenstein, 1986). Within the context of science, argued Cohen, 'facts' and the laws or theories which explain and order them are utterly interdependent. A fact is not merely a datum which enters the body of science from the outside; a fact is rather the sum-total of references to it in the various laws and theories, together with an unexplained and, therefore, problematic residue which remains a challenge to science (Funkenstein, 1986; Popper, 1975; O'Hear, 1989).

Scientists, so far, have shown that it is possible to explain some of the mysteries of our lives and our world in the sense of *what* are the causes of certain events and *how* causal processes work. But to my mind it is unlikely that science will ever provide an answer as to *why* reality is as it is perceived. Scientific inquiry that would deal with such ultimate causes would arrogate to itself the ability to understand the *why* of the very existence of our universe and would claim a perfect and total comprehension of what the universe is, something that, I believe, science will never attain.

NOTES

1 Advanced techniques of artificial insemination and *in vitro* fertilization (test-tube conception) are also processes in which chance plays a part. Even in genetic engineering there are no 100% yields in the scientist's experiments.

CHAPTER FOUR

CHANCE, IMAGINATION, DECEPTION

Science is often imagined as a faultless, giant, monolithic system of knowledge production that uses logical thinking and is aided by the precision of measurement techniques and sophisticated experiments; the scientific ideas and results being expressed by mathematical formulae—a kind of code language inaccessible to the non-scientist. Most scientists also seem to believe fervently that, despite some temporary difficulties, science marches on toward eventual conquests and complete success, seen by the definite answers it can give to practical problems, as well as to the more basic questions concerning human existence and the existence of the universe. Science can, it is argued, come up with perfect or near-perfect predictions concerning innumerable occurrences in reality.

Such views are based on traditional ideas concerning the objectives of science. As Ben-Ami Lipetz notes with reference to such objectives: "one can frequently find assertions that it is the ultimate objective of science to discover something absolute, such as 'truth', or immutable 'laws of nature', or 'the pattern of the universe', or 'the go of things'. The central idea in assertions of this kind is that there is order in nature—that nature is subject to laws of behavior which are invariant with time or place and independent of man; the goal of science is to recognize these laws and to state them for Humanity" (Lipetz, 1965). No less a person than Sir George Thomson, a Cambridge scientist, Fellow of the Royal Society and a Nobel Laureate, in considering the question of what science is trying to do, asserted that "Science is essentially a search for truth" (Thomson, 1961).

It is not my intention to discuss here the highly philosophical issue of what is truth—whether it is conformity or correspondence with facts, an agreement with reality, or a single complete and coherent system which makes it into something final and absolute. Famous thinkers since the seventeenth century have expressed more and more doubts about the idea that science can lead to absolute truth, or that scientific logic can prove such truth. Yet most of the same thinkers, including contemporary scientists and philosophers of science, seem to believe that reality does have absolute laws and unchallengeable truths, and that science is trying to discover them. Why is there such a hankering after a solid base of absolute laws, final truths and perfect knowledge? And why is there even a belief, among many scientists today, that science is edging progressively nearer such knowledge? After all, the overwhelming evidence suggests that despite science's many achievements, the final absolute target seems to be moving further away; science does not seem to be able to catch up with it! Thus, a colossal amount of energy and money has been spent on scientific research aimed at relieving mankind of the scourge of cancer, resulting in some successes, yet without obtaining basic knowledge of how to eliminate the disease. As if to add a further challenge to science in the medical field, AIDS (autoimmune deficiency syndrome) has appeared on the scene in the last two decades. Scientists are not only baffled by the HIV (human immunodeficiency virus, which causes AIDS), but also fear that mutations in the virus could spread the disease through simple contact, leading to mass epidemics of frightening proportions.

It appears that despite the failures of science, the unsolved problems, the chance discoveries, and the constant discrediting and rejection of earlier scientific theories which were once generally accepted, scientists still believe that they are working to reveal the underlying truths of nature. It seems that such belief is both comforting for the scientist's mind (Lipetz, 1965), and also provides that elan or driving force which scientists need in their highly challenging work. With no intention of discouraging or belittling the aims of science, I shall turn in the next sections to the facets of science which show the importance of chance findings, the erroneous assumptions and theories held in the not very distant past, the failed experiments, the interplay of science, chance and personalities, the interweaving of science with unfounded speculation (see Chapter 5), and the unfortunate occurrence of self-deception and fraud. This side of the coin will give us quite a different impression of science, compared with the image of the unswerving, logical, all-knowing, successful and powerful enterprise that it is supposed to be.

The Chance Factor in Breakthroughs

Let us now look at some examples that will highlight for us the role that *chance* has played in some important scientific discoveries. There is a certain abruptness involved in such discoveries. The scientist's 'flash of imagination' is something instantaneous which allows him to observe the chance phenomenon, a fortuitous occurrence, to recognize it as something that deviates from what was known hitherto, and to instantly come to a new conclusion. Yet it is equally true that without the historical background of scientific development, without the training and the experiments of the scientists, and without the accumulation of knowledge and the cooperation of other scientists, fully successful and applicable discoveries would not materialize out of such chance occurrences. Nevertheless, we concur with R. Taton that: "A certain number of examples of accidental observations leading to fundamental discoveries has given credence to the opinion that chance plays a primary role in scientific creation, and that many discoveries would not have been made without the fortunate occurrence of very exceptional circumstances, without which any efforts on the part of the research workers would have remained unfruitful or at least very unrewarding" (Taton, 1957).

Our first example of the role that chance plays in scientific research is the oft-quoted account of Louis Pasteur's (1822–1895) experiments, which established clearly the principle of immunization and led to its practice against infectious diseases. Of particular interest is his experimentation on fowl cholera in which the accidental chance factor occurred. Of course, some knowledge which already existed, and Pasteur's own research gave him the insight to recognize the importance of the chance occurrence. Since the early part of the 17th century it was thought that certain diseases were due to the invasion of the body by some alien matter. But it was only after some two hundred years that the idea of vaccination was accepted; cowpox and human smallpox being regarded as the same disease (indeed, the word vaccine comes from *vacca*, the cow). By the middle of the 19th century there was growing evidence that infectious diseases were due to the presence of micro-organisms. Pasteur's research and the 'germ' theory of fermentation clearly pointed in this direction. There was also the recognition that there were different degrees of virulence in the microscopic agents and that a distinction should be made between microbial and viral diseases: those due to bacteria which are microbial or of the microscopic organism type, as against those due to viruses, which are sub-microscopic agents. It was also known that immunity could be gained by those who survived a disease like smallpox; they would not contract the disease again.

Now let us return to Pasteur's experiments. He pursued his idea that both fermentation and disease were caused by micro-organisms. In his experiments he tried to isolate the micro-organism in a pure 'culture', the latter being the term for a laboratory preparation of a micro-organism. He tried to prove this by producing a culture of chicken cholera, which he intended to inject into fowls and thus show that the microbes caused the disease. In 1879 Pasteur went on a summer holiday and left behind such cultures in his laboratory. On his return he encountered an unexpected obstacle. The cultures of the bacteria he developed had become sterile. He tried to revive them by means of a fluid medium, and then inoculate the resultant subculture into fowls. This did not work; the subculture failed to grow and the birds were not affected. He was aware of the role of oxygen in fermentation and thought of the possibility that the length of time that cultures remained without renewal of microbes or the medium would affect their virulence. He was about to discard everything and start afresh when he had the inspiration of re-inoculating the same fowls with a fresh culture. The result was most surprising: nearly all these fowls withstood the new inoculation, although a new batch of fowls which were inoculated were affected after the usual incubation period. Thus, the principle of immunization with 'attenuated pathogens', that is to say with weakened bacteria, came to be recognized (Conant, 1947; Harre, 1983).

It is interesting to note that in announcing his discovery Pasteur did not clearly mention the accident which was so vital to his discovery. Instead, he stated that, by simply changing the process of cultivation of the parasite; by merely placing a longer interval of time between successive seminations, we have obtained a method for decreasing virulence progressively, and finally get a vaccinal virus[1] which gives rise to a mild disease, and preserves from the deadly disease. But he did make an allusion to the accident when he stated that "Chance favours only the prepared mind" (Harre, 1983). The idea of using attenuated pathogens was successfully applied in the case of poliomyelitis, a paralytic disease which rose to epidemic proportions in the first half of this century. It was Jonas Salk, the American scientist, who used a killed polio virus to develop the first polio vaccine in the 1950s and thus eradicated this disease.

The discovery of penicillin by Sir Alexander Fleming in 1928 further illustrates the role played by accidental happenings and chance observations by scientists, which led to the great therapeutic achievements of this century—the use of antibiotics, which has saved hundreds of millions of lives. Fleming, who was undertaking experiments, noticed on one occasion that one of his cultures became contaminated by a fungous micro-organism from the air outside. As R. Taton tells us, "Such a fact in itself was nothing but an insignificant accident, a small misadventure happening frequently to all research workers in laboratories with insufficient funds to take all the necessary precautions to prevent such contamination. But instead of neglecting this incident and continuing his work on other preparations Fleming went to the trouble of observing the contaminated plate in greater detail." Fleming was rewarded for his effort, for he came upon a very surprising phenomenon: the preparation of bacteria that had been attacked by the microscopic fungi "had become transparent in a large region around the initial zone of contamination." He came to the early conclusion that "this effect could only be due to an anti-bacterial substance secreted by the foreign micro-organism and then spreading into the supporting medium of the preparation." Fleming then decided to study more systematically the properties of this secretion, which had so strong an antibiotic effect (that is, toxic to other species of bacterial organisms). Thus, departing from his plan of research, the English biologist began to study the principal properties of the anti-bacterial substance secreted by the fungi that had accidentally contaminated his plate (Taton, 1957).

Of course the full benefit of Fleming's discovery, to be seen in the successful antibiotic treatment of numerous diseases, such as tuberculosis, diphtheria, typhoid, pneumonia and many other infections, was due to the subsequent concerted and well-planned research efforts of many scientists, the most prominent being the 'Oxford Team' led by E.B. Chain and H.W. Florey. The reader may think that the stories of immunization and antibiotics simply depict an historical development from poorly funded and individually executed research, fraught with accidents and unexpected discoveries, to well-planned modern medical research supported by the necessary resources and executed by teams of scientists bent on achieving their projected aims. Some such change has, of course, taken place, yet chance remained a factor in modern research for a long time. Alexander Kohn tells us that: "The role of chance in the discovery of most of the modern drugs is very significant. Indeed, there are very few drugs that were designed with the prior knowledge of exactly how and where they would act" (Kohn, 1989). The elements of luck and chance in medical research and in the development of drugs were important until about the middle of this century, but since then their importance has certainly declined.

From the field of physics, too, a number of examples can be brought to show the importance of serendipity in science. There are instances where scientists "set out to study one phenomenon and in the course of their experiments stumbled upon a new manifestation" (Ashall, 1994). One such case is that of the German scientist W.K. Röntgen (1845–1923) who discovered X-rays by chance observation, in 1895, while experimenting with electric current flows. Another example can be given in connection with the discovery of radioactivity in 1896 by A.H. Becquerel (1852–1909) in France. Following Röntgen's discovery of X-rays, Becquerel began to look for X-rays in the fluorescence observed when certain salts absorbed ultraviolet radiation from sunlight. By chance Becquerel discovered that the salt gave off a penetrating radiation independently, without ultraviolet radiation. He deduced that radiation came from the uranium in the salt. In Bernal's account, what actually happened can be regarded as "something like a real accident in the history of science." In the first place it was merely a hint by Henri Poincaré (1854–1912) that had caused Becquerel to seek to ascertain whether there was any connection between X-rays and phosphorescence. But more importantly, Becquerel, who used his father's magnificent collection of phosphorescent substances, "might just as easily have picked on zinc sulphide as on uranium nitrate and the discovery of the phenomenon of radioactivity and all it meant for atomic physics might have been delayed for another fifty years. Who knows how many equally simple phenomena capable of revolutionizing our science now lie hidden around us?" (Bernal, 1965).

Of course, serendipity is usually coupled with intentional scientific research; and the training and knowledge of the scientist is necessary, since only he can appreciate the importance of his chance findings. At the same time new chance findings often induce further tremendous progress in science. A case in point is that within a very short time of the discovery of radioactivity, Pierre (1859–1906) and Marie (1867–1934) Curie, the well known French and Polish scientists, had laid bare the essential features of *spontaneous atomic change.* Further research by Ernest Rutherford (1871–1937) and Frederick Soddy (1877–1956) added to the understanding of the nature of the break-up, disintegration and transmutation of atoms (Bernal, 1965; Harre, 1983).

The work of the scientist is not only often marked by accidental discoveries, but also by failed experiments. However, the single-minded scientist often succeeds despite such failures. Thus, after many difficulties Michael Faraday (1791–1867) succeeded in inventing a new technology based on a new scientific concept. He not only discovered how to obtain electricity from magnetism, but he also elaborated a descriptive theory as to how it occurred. He applied his theory of electrical action, and after some failed experiments, he invented the first crude dynamo, as he produced a continued current of electricity by moving a wire conductor to and fro between the poles of an electro-magnet. Faraday defined the laws of electro-magnetism in terms of the *relative motion* between lines of magnetic force and conductors. Thus his theory was a precursor of the theory of relativity (Crowther, 1935).

The examples above serve the purpose of showing that chance plays a definite role in scientific discoveries. Continuing with R. Taton's excellent treatment of this subject, we must distinguish between "psychological chance", that is, "the fortuitous coming together of two ideas; which as it were...is a constant factor in every discovery"; and "external chance", that is to say, "accidental facts which in some cases have led research workers into particularly fruitful directions". It is seldom that a discovery is produced by purely "external chance", although the latter could be of great importance. Much more frequently it is "psychological chance" that is at work, where the apparently fortuitous invention, as in the research of Pasteur or Fleming, is in fact based on a great deal of preliminary work and "is fashioned by the intuition (and aesthetic sense) of a particular research worker" (Taton, 1957). "Psychological chance" occurs within the wider realm of *scientific imagination,* which quite obviously was at work in the case of Rutherford. As Peter Medawar says: "Every discovery, every enlargement of the understanding, begins as an imaginative preconception of what the truth might be" (Medawar, 1984). I would change this to 'what the best approximation of reality might be'. Science pursues this aim by its observations, experiments, discoveries, hypotheses and theories. In all this "the generative act in science is the brainwave, inspiration or flash of imaginative insight" that appears in the form of an hypothesis. William Whewell called them "happy guesses" and "felicitous strokes of inventive talent". There is no doubt that scientific ideas, appearing as hypotheses before or after experiments, observations or the gathering of data and their analysis, and the subsequent testing of such hypotheses for the purpose of verification or falsification, and their eventual emergence as full-fledged theories, are the products of imaginative thinking. In this sense they are the result of the creative and inventive faculty of the researcher, of his ability to conceive what reality is like (Medawar, 1984; Perutz, 1989). This constant creativity is seen also in the experimentalist definition of science, by James Conant: "that portion of accumulative knowledge in which new concepts are continuously developing from experiment and observation and lead to further experimentation and observation" (Conant, 1947).

Erroneous Assumptions, Misinterpretations and Failed Experiments

Scientific imagination can, of course, run riot and produce untenable theories. Similarly, excessive faith in experimentation can mislead the scientist and produce dead-end situations, seen in the fact that scientists undertaking the same kind of experiments sometimes obtain contradictory results. How do such scientific failures occur?

To answer this question let us first consider the ill-fated *phlogiston theory* (from the Greek, 'inflammable'), which I mentioned briefly in Chapter 1. G.E. Stahl (1660–1734) and later Joseph Priestley (1733–1804) put forward this theory which seemed to make sense in the science of chemistry; a theory of combustion which was accepted during the 18th century until in 1772 A.L. Lavoisier refuted it. It may be noted that although Priestley was one of the discoverers of oxygen, to which he pointed as something that green plants gave off and which he called "dephlogisticated air", and although he cooperated with

Lavoisier, he held on to his theory according to which all combustible substances were composed of phlogiston, which escaped upon burning, and ash that remained. The phlogiston theory, like any other scientific theory, was based on observation and assumption, and it involved some reasoning. Since any material that burned changed its character after the burning, Stahl, for instance, assumed that part of the material must have escaped, that was the burning part, and that this was a substance called phlogiston, "which deserts the burning mass to leave behind a formless pile of ashes" (Gillispie, 1960). But both his assumption and the reasoning were wrong. A.L. Lavoisier, who based his understanding on his own experiments, and who provided the correct reasoning and the accepted theory in science today, wrote; "I discovered that sulphur, in burning, far from losing weight, on the contrary gains it:....this increase in weight arises from a prodigious quantity of air that is fixed during the combustion and combines with vapours". It was obvious that, had something escaped, the remaining material after the burning should have been lighter in weight. Lavoisier was convinced that both combustion and respiration involve a combination with something atmospheric, and pointed to oxygen as the element which combined with any material that underwent the process of burning. In general, where in fact there was a gain in oxygen, Stahl saw a loss in phlogiston (Gillispie, 1960). The theory was, of course, discredited; something that often happens in science.

Another theoretical assumption that was discredited at the end of the last century was that of the *ether*, which for a long time scientists believed existed throughout space. Newton himself thought that gravitation was transmitted through an ether composed of imponderable atoms, forming an element far rarer, subtler and much more elastic than air. Similarly Immanuel Kant, the greatest philosopher of modern times, in an unfinished manuscript written at the turn of the 18th and 19th centuries suggested that the study of ether was necessary since, in his view, action at a distance was not possible. In other words, there had to be some medium in space through which things moved; otherwise how could, for instance, light be transmitted? The idea of *vacuum* seemed unacceptable to the scientist and philosopher. Therefore, it was believed that a substance called 'ether' actually permeated all matter and space, a substance that was assumed to be weightless, transparent, frictionless and undetectable. We also find that James Maxwell (1831–1879) evoked the image of the "great ocean of aether" (Gillispie, 1960). Karl Pearson (1857–1936) in his book *The Grammar of Science*, first published in 1892, expressed his doubts about the existence of ether, and thought that "like many other conceptions of which we have no direct perception", ether is just a conception which "physicists project....into the phenomenal world and assert its real existence" (Pearson, 1937 edition; Toulmin and Goodfield, 1961; Kockelmans, 1968).

But some physicists set out to test the existence of ether. In their famous experiments, during the 1880s, A.A. Michelson and E.W. Morley reasoned that if, indeed, the earth moving in its orbit around the sun (at about 30 kilometers a second) passed through this supposed element of ether, then their measurements of the speed of light in the direction of the earth's motion as compared to the speed of light at right-angles to the earth's motion, should show a difference. No such difference was found. This not only weakened the concept of ether, but it also paved the way for Albert Einstein's proposal in 1905 that the speed of light is a universal constant (see Chapter 2, pp. 34-35). Eventually, as a result of the Michelson-Morley experiments, the very idea of the existence of an ethereal element in space was totally discarded (Schroeder, 1992).

A most bizarre case of failed experiments and scientific falsification, involving charlatanism and politics, was that of T.D. Lysenko, a Soviet agronomist, who turned out to be a pseudo-scientist who destroyed the modern science of biology in the Soviet Union in the thirty years between 1934 and 1964. Lysenko and his colleagues experimented during the late 1920s with different strains of wheat, and basing themselves on the theories of the French biologist J.B. de Lamarck, maintained that the direct transformation of spring wheat into winter wheat and vice versa was possible through a process called "vernalization". Furthermore, he claimed that the adaptability of the wheat which induced certain new characteristics would be inherited. The theory of "inheritance of acquired characteristics" ran counter to modern genetics, which held that heredity was controlled by genes and, therefore, environmental changes could not be inherited. "There is no molecular mechanism enabling instructions from the environment to be imprinted on DNA directly" (Jacob, 1989). Lysenko was prepared to fly in the face of accepted modern science and to enforce his views at gunpoint, with the blessing of Stalin and Khruschev. Eventually it turned out that the 'data' that Lysenko gathered in collective farms to support his theories were largely false. The resulting crisis in Soviet agriculture and the fall of Khruschev ended this unfortunate episode in the history of science (Bernstein, 1980). The rejection of Lysenkoism does not mean, however, that adaptability to the environment as an existing process is not accepted in modern biology. It is, for instance, accepted that over millions of generations biological species 'evolve'; more specifically, it appears that out of millions of bacteria a very small percentage undergo chance mutations which enable them to adapt to the antibiotic-treated environment, and thus survive medicines aimed at killing them, and multiply. Such theories and facts are subjected to objective scientific scrutiny, seen for instance in the Human Genome Project. The latter is a 15-year global effort, started in 1990, aimed at discovering all the 60,000 to 80,000 human genes, and making them accessible to further biological study. Such mapping of the entire human genetic material should ultimately determine the complete sequencing of DNA subunits, and should also help in the study of genetic disorders, mutations, and cloning (McKusick, 1989).

A far less serious case, one of self-deception, was that of the famous American scientist Linus Pauling, a double Nobel Prize winner. Pauling was awarded the Nobel Peace Prize for his campaign in the 1950s and 1960s against nuclear tests, and the Nobel Prize in Chemistry for his pioneering work in the 1920s and 1930s on the structure of molecules. At one time he was ranked alongside the greatest scientists of all time. He died in 1994, having endured the derision of his scientist colleagues for the last two decades of his life because of his strong belief, without scientific foundations, in the health benefits of Vitamin C, an organic substance found in food such as citrus fruits and vegetables. Pauling maintained that large doses of this vitamin can help prevent or treat almost any disease—the common cold, even cancer and AIDS. Other scientists, however, duplicated Pauling's research and concluded, for instance, that Vitamin C was useless in treating cancer. Such were the results of independent tests of the highly respected Mayo Clinic in Rochester Minnesota, in 1979 (Reuter, 1994).

Cases of self-deception in science have been numerous. Following the most important discovery of X-rays by Röntgen in 1895, a leading French physicist René Blondlot announced in 1903 that he discovered another sort of rays, which he named N-rays after the University of

Nancy in France, where Blondlot's experiments took place. The main claim of this scientist was that "when N-rays impinged on an object....there was a slight increase of brightness". Blondlot's warning that "the general properties of N-rays were elusive" and that "a great deal of skill was needed to see the effect of these rays" should have alerted scientists immediately as to the doubtful basis on which this scientific assertion was made. Yet, many other scientists reported seeing the N-rays; they were in fact victims of self-deception, probably because they thought that considering the importance of the discovery of X-rays, this new type of rays was similarly likely to be of importance. Also in the case of many French scientists, national pride added to their insistence that N-rays exist. In fact within a short time the scientific community concluded that such rays did not exist (Huizenga, 1992; Kohn, 1986).

Perhaps the longest lasting scientific misinterpretations and the most embarrassing blunders have occurred in the field of evolutionary biology, which was highlighted by a major fraud that misled scientists for forty years—between 1912 and 1953. This occurred among evolutionary scientists who "have a dismal record when it comes to interpreting hominid fossils". Examples are numerous, but the most famous case is that of the *Piltdown Man*. Hominid fossils are remains of extinct man-like creatures of various species, which according to evolutionary history belong to one 'family', there being now only one species in existence, that of *Homo sapiens,* while all other species, such as *Homo erectus*, regarded as the direct ancestor of modern *Homo sapiens*, being extinct. The theory held that some of the species had certain apelike characteristics and should be regarded as transitional species, which eventually evolved into other species, and culminated in the birth of the species of *Homo sapiens*.

Now let us return to the story of Piltdown Man. In 1912 Charles Dawson, a fossil collector, announced that he had found in a gravel formation on Piltdown Common, on the Sussex coast of England, fossilized remains of a prehistoric man, namely fragments of a cranium, a jawbone and other specimens. This find came to be known as the Piltdown Man and was accepted by the scientific community as genuine. In fact the Piltdown Man turned out to be an outright forgery. By 1953, "after intensive reexamination, the remains were shown to be skillfully disguised fragments of a quite modern human cranium, the jaw and teeth of an orangutan and the tooth probably of a chimpanzee, all fraudulently introduced into the shallow gravels". Chemical tests revealed that the fragments had been deliberately stained, the bones and teeth were filed down and shaped so as to make them fit to the cranium and to give them a more human look. This artificial creation produced overall an ancient look which was befitting a prehistoric fossil (*The New Encyclopaedia Britannica, Micropaedia*, 1992; Aviezer, 1996). So long as Piltdown Man was accorded a high antiquity it appeared as a feasible ancestor of modern man. Most paleontologists, biologists and anthropologists accepted this supporting evidence for Darwin's theory of evolution. It is claimed, however, that some anthropologists never accepted the seductively plausible idea that Piltdown Man represented the 'missing link' with mixed ape-human features (*Anthropology Today*, 1971).

It is not very important to know who perpetrated this hoax and why. Rather, I agree with Nathan Aviezer's comments that: "What is important here is *not* the fact that a fraud was perpetrated—every profession has its cheats. What is central to our discussion is the assessment of this fraudulent fossil by leading members of the scientific community. One would have thought that as soon as this jaw of an ape reached the hands of the professional anatomists, the game would be up. How could any skilled anatomist fail to recognize that the Piltdown jaw was identical in every respect to that of a modern orangutan and that the Piltdown skull was identical in every respect to that of a contemporary man, without any indication of those features that characterize 'prehistoric man'? Surely, Dawson's fraud would be exposed by any of the leading scientists within a matter of minutes" (Aviezer, 1996).

And yet three leading scientists in British paleontology, specializing in the scientific study of life of the geologic past, confirmed the Piltdown Man as genuine. How could Grafton Elliot Smith and Arthur Keith, two foremost British anatomists, and Arthur Smith Woodward, an expert in hominid paleontology, all knighted for their scientific achievements, make such gross blunders? The answer, Nathan Aviezer tells us, lies in the preconceived notions of the British school of paleontology in those days, which ensured that these scientists saw what they wanted to see. Their theory was that "the brain (skull) of Modern Man should have evolved relatively rapidly, whereas the jaw should have evolved more slowly. Thus, according to the British school, it was to be expected that our prehistoric ancestors would at some early stage have a relatively modern human-like skull while still sporting a relatively primitive ape-like jaw. When Piltdown Man displayed precisely these characteristics, he was welcomed with open arms by the British paleontologists" (Aviezer, 1996).

Britain, however, was not the only country where unfounded 'scientific' assertions were made concerning man's ancestors. Early on this century, in France, Marcellin Boule and, in the United States, Henry Fairfield Osborn made erroneous claims concerning fossil discoveries, all with the view to strengthening evolutionary biology. It appears that these scientists were "motivated by intense personal rivalries and preconceived notions" (Aviezer, 1996; Gould, 1980: Trinkaus, 1993).

The final example is a fossil, known as archaeopteryx, which has been claimed to be a transitional form between reptiles and birds. This "extinct animal of great antiquity" is described in detail, as if it is a matter of scientific fact, without any mention of possible doubts (*The New Encyclopaedia Britannica, Micropaedia*, 1992). Yet it is known that Lee Spetner, a scientist who collaborated with several distinguished scientists including Fred Hoyle, the Cambridge scientist, suspected that this was another case of forgery. There were only two complete specimens in the world, discovered by a father and son in a Bavarian limestone quarry and sold to a museum for a substantial sum of money. Spetner managed to take infra-red photographs of one of the specimens, and he and his collaborators believed after examining the photographs that the fossil was not genuine. Cyril Domb, an internationally well-known scientist, relates how a paper containing their findings was rejected by *all* the biological journals (Domb, 1995; Kohn, 1986). This certainly smacks of unscientific collusion among some scientists strongly entrenched in their views in a particular area of science, where serious doubts were being raised by other scientists. Thus, in the 1980s the Darwinian theory of evolution was being reevaluated. The world famous paleontologist, of the American Museum of Natural History, Niles Eldridge held that: "The pattern that we were told to find for the last one hundred and twenty years does not exist" (Schroeder, 1990). The problem with Charles Darwin's explanation is that if evolution works in small cumulative steps over vast periods of time, with more primitive species gradually evolving into more complex species, this means that if the explanation is to be correct also regarding large-scale changes differentiating one class of animal from another, "then the transitional forms between one type of animal and another must also appear in the fossil record. The fact is that these predicted transitional forms have never been found" (Aviezer, 1990). It may be noted that the method of radiocarbon dating, using the rate of radioactive decay in all matter containing radioactive carbon to calculate the age of fossils, is useful only for dating fossils of up to some forty-thousand years old, but not for dating fossils of greater antiquity.

Some evolutionary scientists, like Richard Dawkins, are most vehement in their stand on the veracity of Darwinian theory despite the lack of fossil evidence. Others, however, even though they support evolutionism, admit that the evidence is very incomplete concerning, for instance, the descent of man from ape-like creatures. Thus Bronowski: "There is now a blank in the fossil record of five to ten million years. Inevitably, the blank hides the most intriguing part of the story, when the hominid line to man is firmly separated from the line to the modern apes. But we have found no unequivocal record of that yet. Then, perhaps five million years ago, we come certainly to the relatives of man" (Bronowski, 1973).

So far, the examples I have brought of erroneous assumptions, self-deceptions and the acceptance by scientists of forged evidence have been from the natural sciences, although in the case of the Piltdown Man both paleontology and anthropology were involved. The social sciences have also their share of errors, misinterpretations and claims based on apparently forged data, as well as of theories influenced by ideological views. I ashall adduce two cases from the social sciences, both of which concern the never-ending issue of 'nature versus nurture', that is, the influence of heredity and environment on human ability and behavior.

The first case is the study conducted by the anthropologist Margaret Mead in the early 1920s on the Pacific island of Samoa. Her topic was adolescence, sex mores and family ties among Samoans, and she based herself mainly on interviews she conducted, via an interpreter, with some 25 adolescent girls over a period of nine months. Her conclusion was that in Samoa a stressless primitive culture prevailed, which led to an easy and conflict-free life for adolescents, whose family ties were casual, whose sexual mores were relaxed, and whose life was devoid of guilt and adolescent turmoil. Her book, *Coming of Age in Samoa*, published in 1928, became very popular in the Western world. It was welcomed especially by those arguing for a more liberal approach to family life.

Yet five years after Mead's death in 1978, Derek Freeman, an Australian anthropology professor, published a book in which he debunked the conclusions drawn in *Coming of Age*. Freeman's view was also upheld by the American anthropologist Brad Shore. Both Freeman and Shore spent years in Samoa and Freeman also learned the language. These critics maintained that Samoan culture was authoritative, that children were under considerable stress, to the point of enduring severe punishment, and that among Samoans violence (homicide and rape) occurred much more frequently than in the Western world. Now, we must consider the possibility, raised by Alexander Kohn, that since a whole generation passed between the studies conducted by Mead and Freeman, "during this period of international influence by radio, and television, some cultural patterns may have changed throughout the world, as well as in Samoa" (Kohn, 1986). It is, however, more plausible that Mead relied on faulty and flimsy evidence gleaned from a couple of dozen young girls who told her stories they believed she wanted to hear, and that she was strongly influenced by her mentor Franz Boas who upheld the doctrine of *cultural determinism*, the view that human behavior is most strongly determined by the cultural environment and very little by genetically determined and inherited traits. It has been argued that Mead's erroneous conceptions and flawed theories were due to her wish to support her ideology; in short "she found what she was looking for" (Freeman, 1983; Mead, 1933; Kohn, 1986).

The second case concerns a prominent British psychologist, Sir Cyril Burt, who was professor at the prestigious University College, London, from the 1930s until 1951. His influence on the educational system in Britain was tremendous, and brought about the selection of children at the age of eleven, by means of IQ testing, into three different educational levels. Basically, he claimed what many other studies also claimed; namely, that intelligence was inherited. In his earlier researches he tested groups of children aged 6–14, and later, in the 1950s and 1960s, he carried out crucial tests on 53 pairs of identical twins. His results showed very high correlations even between the IQs of twins brought up separately and in different environments. Some of Cyril Burt's statistical methodology and findings were questioned during his lifetime, but only after his death in

1971 did a sufficient number of converging facts come to light to cast serious doubts on the reliability of his data. Worse still, some of the facts convinced investigative journalists as well as statisticians and social scientists that much of his data were simply concocted. Briefly, the following are the reasons for such an accusation: (a) the variation in his results was below the likely vagaries of chance; (b) fuller accounts of sources, calculations and detailed tables, which Burt claimed were contained in students' degree theses, were never found, since they were not filed with the university; (c) two women collaborators, in crucial papers published in the 1950s, could not be traced; (d) the same women's names were, apparently, used as pseudonyms for reviews praising Burt's publications, in a prestigious journal which he himself edited; (e) the raw data of his work (test sheets and notebooks) could not be inspected since these were burned. By the 1970s even some of his supporters admitted that Burt may have regarded the actual data as "merely an incidental backdrop" for the illustration of the theories he held important, the implication being that he might, indeed, have invented his data. Many others would simply classify Burt's acts as 'scientific fraud' (Kamin, 1974; Wade, 1976; Kohn, 1986).

Returning to the natural sciences, a more recent extraordinary scientific claim, which soon became disputed and which has not stood up to the scrutiny of the wider scientific community, is that of cold nuclear fusion. John Huizenga has described this as "the scientific fiasco of the century". Both nuclear fission, in which the atomic nucleus splits, and nuclear fusion, in which the joining of light atomic nuclei produces a heavier nucleus, release very large quantities of energy. Nuclear weapons have derived colossal explosive power from nuclear fission or a combination of nuclear fission and nuclear fusion. Scientists with the desire to exploit nature's processes for peaceful purposes, have been trying for many years to find a way of tapping nuclear fusion's potential as a commercial source of energy. We know that it is fusion that powers the sun and other stars. But on earth, atomic nuclei normally stay too far apart to fuse. Their proton particles have positive electric charges and the mutual repulsion of like charges keeps the nuclei separated. Since the 1950s, scientists have tried to overcome this problem by heating hydrogen isotopes[2] to temperatures of tens of millions of degrees. At such high temperatures, atoms move so swiftly that they get close enough to fuse. However, the high temperature causes the nuclear fuel to expand with explosive force, thus cutting off the fusion reaction before usable amounts of energy can be produced (*Science Year*, 1991).

The tremendous cost of high temperature fusion, and the inability to produce cheap energy by this method, were the factors that prompted scientists to turn their attention to cold fusion.[3] On several occasions during this century there were claims of success which turned out to be unfounded. The most recent claim by electro-chemists Martin Fleischman and B. Stanley Pons, at the University of Utah in Salt Lake City in the Untied States, was the most spectacular. In March 1989 it was announced at a press conference that these scientists had successfully produced nuclear fusion in a simple table-top device at room temperature by a very easy method. More precisely, the device is a jar containing two electric terminals and filled with 'heavy water' (that is, the isotope of hydrogen, termed deuterium). Fleischman and Pons claimed that when they passed a current from one terminal to another a large amount of excess heat was generated (four watts of output power for each watt of input power), and there was an emission of neutron and gamma rays (particles and radiation from the excited atomic nuclei), these being tell-tale signs of nuclear fusion. They asserted that the pressures attained during electrolysis (the chemical decomposition of substances by an electric current passed through the substance in a dissolved state) were sufficient to drive deuterium nuclei close enough to fuse. That is to say, the researchers held that the atoms of heavy isotopes trapped in the terminals had been crowded close enough together to cause a sustained nuclear fusion reaction (*Science Year*, 1991; Huizenga, 1992).

A press release on behalf of the scientists claimed that "the discovery will be relatively easy to make into a usable technology for generating heat and power", and that this breakthrough will provide humanity with an unlimited supply of energy, since the electro-chemical techniques used fuse some of the components of heavy water which contains deuterium (the isotope of hydrogen) which occurs naturally in sea water. This energy source is virtually limitless and, as stated, obviously superior to that of the nuclear fission process, used today in nuclear power plants, since it creates a minimum of radioactive waste, and is superior also to other energy sources such as coal, gas and oil, which pollute the environment and which are bound to become depleted (Huizenga, 1992). These announcements "sent chemists, physicists and other scientists world-wide scurrying to their laboratories to try to reproduce the Utah phenomenon. ... Quite uncharacteristic of scientific research all over the world very early details about the Utah experiments were sketchy at best and scientists had to work mainly from accounts in the popular press. When [a scientific] paper was finally available for examination by an anxious scientific community, most readers were shocked by blatant errors, curious lack of important experimental detail and other obvious deficiencies and inconsistencies" (Huizenga, 1992).

Soon after the announcement of cold fusion, the claimed discovery was presented and debated at scientific conferences, and scientists tried to duplicate the experiment in other laboratories. In fact a review process took place which soon showed emphatically that the claim was incorrect. The crux of the debate was whether substantial excess heat (energy) could actually be produced by the Fleischman-Pons method. Scientists who rejected this claim argued that the commensurate intensities of fusion by-products, the crucial part of the signature of nuclear fusion, have not been found. It is interesting to note that although the vast majority of scientists were unable to replicate the cold fusion results, over a hundred groups reported excess heat (energy) and/or fragmentary evidence for some fusion by-products. How is it that these positive claims occurred?

The answer to this question must be divided into two parts. First, the proponents of cold fusion targeted the 'science establishment' as responsible for actions detrimental to the growth and development of cold fusion; cold fusion had attractive features, in that the experiment appeared to be very simple, and required only elementary equipment; there was enthusiasm for its potential usefulness; the so-called discovery fulfilled the hopes of an energy-hungry world, and encouraged those who stood to profit financially. All these factors produced a favorable climate, encouraged the believers and peaked in a 'fusion frenzy'. Moreover, there were claims that working on such a sensitive subject required some degree of secrecy; certain shortcomings in the instrumentation were blamed for the difficulty of proper detection and some incorrect results; the rush to produce results was blamed for some lack of detail or proper pre-publication peer review of papers; there were even some theoretical papers which explained the phenomenon despite the fact that such explanations contradicted the accepted nuclear physics theory. The enumerated factors were conducive to continuing belief in and reporting of cold fusion by many scientists besides Fleischman and Pons.

However, almost from the start there were inconsistencies in experimental results. For example, a group at Brigham Young University attempting to reproduce the Fleischman/Pons results did not detect excess heat; their detection of possible fusion by-products showed such a low intensity that these could be ascribed to background effects whose sources were cosmic rays or natural radioactivity. Skepticism among scientists increased as time went on and the number of negative results was becoming overwhelming. In short, there was no clear evidence that the claimed excess heat, if real, is associated with a nuclear process, and the theories that tried to explain the phenomenon were not verified. Today, years later, the incorrectness of the claim is shown clearly by the fact that the process has not been applied to the production of usable energy anywhere in the world.

The cold fusion fiasco and other scientific self-deceptions have been held up by many scientists as examples of bad or pathological science, defined by Irving Langmuir as "the science of things that aren't so" (quoted by Huizenga, 1992). Sociologists of science will ponder on how social factors such as competitive drives, institutional and national loyalties, the promise of fame and financial gain, produce a lack of realism, serious mistakes and false claims in the scientific enterprise; and how disagreements result in the categorization of scientists, regarding some disputed claims, into believers, doubters and nay-sayers. The examples are too numerous to call them rare aberrations. However, in the long run the scientific process seems to be able to correct its own errors (Huizenga, 1992). After all, the scientific community has not accepted any of the strange or extraordinary claims which through scientific checking are declared to be erroneous.

Science as Is

To conclude this chapter let me, first of all, draw attention again to the chance factor, which sometimes plays a positive role in a new discovery, yet at other times it prevents the scientist from succeeding in his work by leading him up blind alleys. In the best recorded cases, the chance factor in scientific discoveries has been, most certainly, a boon—as in Louis Pasteur's immunization against infectious diseases, or Alexander Fleming's discovery of antibiotics. This is not to say, however, that such discoveries are purely accidental events. The scientist in his observations is often on the look-out for novel phenomena, and much depends on his scientific background, his personal abilities, as well as on his flash of insight or sudden inspiration at the time he makes his observations. In this connection it is interesting to note Alexander Kohn's comments on missed opportunities in science. Here, too, chance and personal factors mix to affect the outcome: "Among those who could have made a discovery, but did not, are many who missed the opportunity, either because they did not fully appreciate the importance of their results, or because they were too timid to insist on the correct, but novel interpretation of their findings which ran against the accepted paradigm. Missed opportunities are also the result of the background, the education and the environment of the scientist as well as of his or her preoccupations, bias and goals" (Kohn, 1989).

One can present the importance of chance yet from another perspective. A scientist has a chain of thought which could lead him to a certain hypothesis to be tested by experiment. But suppose that he is having second thoughts, which lead him in a different direction, or that by 'associative' thought he 'finds' what he regards as the 'missing link', and thus arrives at a different hypothesis, and in turn the new hypothesis leads to a discovery; can we say in such cases that the 'second thoughts' or the 'association of ideas' occurred by chance and, therefore, the discovery was due to chance? The answer seems to me to be *yes*. But this does not mean that science progresses by sheer luck, flashes of insight and imagination, or by serendipity. There are other factors equally important which are at work. There is a process of problem solving, a logical step-by-step procedure which is essential in science. And as a matter of fact, the very logic applied and the style of thought used by scientists are what sociologists would recognize as cultural products of interaction of scientists in a collective, that is the scientific community of which they are part (Kohn, 1989). In the end, it is a combination of chance, imagination, logics, carefully designed experiments, the scientist's personality and training, and the influence of the scientific community, which produce scientific results. It is true that scientific knowledge emanates basically from the interaction of the scientist with the world of nature, yet in many respects the production and appraisal of scientific knowledge is achieved by a complex endeavor (Ravetz, 1971; Chalmers, 1976).

Continuing to sum up our presentation of the human face of science, let me recall that we encountered certain 'human errors', seen in erroneous assumptions, such as that of the ethereal medium in the universe, or misinterpretations such as the phlogiston theory. But some misinterpretations, as that of adolescent life in Samoa may not have been simply a naive human error; it seemed that Margaret Mead had been influenced by the doctrine of cultural determinism. In some cases of failed experiments, more than human error was at work: in the case of Blondlot's N-rays or Pons' and Fleischman's cold fusion, elements of self-deception, personal ambition, the search for fame and money, and national pride, were apparently involved. In other cases, such as Linus Pauling's belief in the effectiveness of Vitamin C for curing disease, the factor that worked may have been sheer obsession. We also came across far worse cases: the charlatanism and political maneuvering of T.D. Lysenko, the forgery of the Piltdown Man and its acceptance as genuine by dogmatically entrenched evolutionists, and the seemingly fabricated data of Cyril Burt. In many of these cases the accepted norms of science, namely scientific honesty and reliance on genuine data, were clearly contravened. The end result, of course, has always been the repudiation and the discarding of the 'faulty' results of science.

My account of certain blunders in science and the fraudulent practices of some scientists, although based on a selection of cases which represent a very small proportion of the vast scientific enterprise, does provide a view of the darker side of that enterprise. The reader should not regard this, however, as an intention on my part to disparage science or scientists as a whole. The achievements of the latter are legion, and the honesty, steadfastness and hard work of the vast majority of scientists is common knowledge. Yet I have chosen to highlight the accidental and negative facets of science in order to deflate the myth of scientific perfection. In giving examples of chance discoveries, mistaken notions, failed experiments, and in mentioning certain foibles of scientists such as self-deception, over-zealous competition, personal and national pride[4], and the use of fraudulent evidence[5], I have followed the scientific principle of taking *all relevant facts* into account. It would be wrong to dismiss some of these facts, unexpected and unpleasant as they may be, as the insignificant hiccups of the scientific enterprise. I preferred, therefore, to present a picture of 'science as is', rather than to produce an unrealistic and idealized model of science.

Notes

[1] The term 'virus' was used in a generalized way to include any disease-causing agent.

[2] Types of atoms of an element which have identical chemical properties but differ in mass because their nuclei contain different numbers of particles.

[3] In fact, cold fusion research started much earlier, in the 1920s.

[4] As an example, see the case of Werner Heisenberg, the renowned German physicist (Rose, 1998).

[5] See also: "documented cheating in clinical and basic research" (Kohn, 1986).

CHAPTER FIVE

THE COSMOS—FROM EVIDENCE TO SPECULATION

Science is an enterprise which is never satisfied with existing knowledge. It is, therefore, continually producing new hypotheses, the testing of which requires hard evidence. This task is particularly daunting in a science such as cosmology, which studies the nature, origin and history of the universe. Here evidence must be sought from the very distant reaches of the universe, both spatially and temporally.

Milton Munitz describes the universe as: "the comprehensive whole that includes all physical objects, processes and events within its scope." Following Paul Davies, we can refer to the universe as an all-embracing concept of a complex, coherent, unified and consistent whole, displaying certain regularities and forces which follow definite physical laws. Munitz also enumerates a number of basic questions asked in recent scientific cosmology: one such question is whether the universe had a beginning, or whether it was always in existence? Another question is whether the universe has a spatial limit, or is it of infinite spatial extent? (Munitz, 1986; Davies, 1992). In this chapter I shall relate to the answers cosmology has provided to these questions and the extent to which these answers are complete and buttressed by scientific evidence.

In order to answer such questions the cosmologist propounds a theoretical model which has to confront the observational data, leading to an interplay between the empirical and its conceptual interpretation. This is the generally accepted scientific process. The data available to the cosmologist is information he receives "in the form of signals conveyed by electromagnetic radiation of various frequencies....obtained by means of optical and radio telescopes, infrared detectors, radio antennas, and satellite mounted telescopes." To give some idea of how far into space and time advanced data-gathering instruments reach, let us note that radio-telescopes have probed enormous distances of billions of light-years, a light-year being "the distance the light, travelling at the rate of 300,000 kilometers per second, traverses in a year." The vastness of the universe may be comprehended if we think of the giant clusters of galaxies of which it is composed, a galaxy being a collection of stars held together, so to speak, by gravitational attraction. The sun is such a star, one among many stars in our own galaxy, the Milky Way, which is itself a member of a local group of galaxies. Our own group of galaxies is a relatively small cluster numbering approximately twenty galaxies—some clusters have thousands of member galaxies. Such clusters or super-systems have also been described as gigantic "cloud complexes" of galaxies (de Vaucouleurs, 1957). The galaxies within these vast clusters or super-systems are separated from each other by enormous distances, and the size of a cluster may be expressed in terms of millions of years that light takes to traverse it, the largest clusters being estimated at 100 million light-years across.

In what ways is this vastness that we call the *universe*, observable and intelligible from a scientific point of view? Galactic astronomy and astro-physics have certainly widened our horizons and deepened our scientific understanding of certain aspects of the universe. Yet, as we shall see, the basic questions of its origin, its extent and its future, remain unanswered by science. Since evidence is our main scientific criterion, let me enumerate, in the simplest terms, some of the major steps along the road to a better understanding of the cosmos. The idea that the universe is composed of many galaxies like our own was already proposed by Immanuel Kant in 1755. The idea became generally accepted by the beginning of the 19th century. But it was not until 1923 that it became scientifically established by the evidence that the US astronomer Edwin Hubble was able to provide with his new powerful telescope. By 1929 the Newtonian view of the universe as infinite and unchanging had to be abandoned once Edwin Hubble discovered that in the spectrum of light emitted by the stars of the galaxies there was a shift toward the red. To understand the meaning of this we must refer to the Doppler effect. In 1842 J.C. Doppler, a Prague mathematician, described a phenomenon that is true of all transmissions of energy. The wavelike characteristics of sound and light which are emitted from a moving source undergo a shift in their frequency relative to a stationary receiver. Thus, an ambulance siren becomes shriller as it comes nearer to where we stand and it becomes less high-pitched as it moves away, since in our vicinity the sound wave travels a shorter distance to us and thus at an increased frequency, resulting in the high-pitched sound of the ambulance siren. The same is true of light waves. A spectrometer picking up the different color bands of light waves will show that light emitted by an object moving towards us shifts to the blue end of the color spectrum, the blue representing a shorter wavelength, hence a higher frequency. On the other hand, light emitted from an object moving away from us shifts to the red, that is, to a longer wavelength representing a lower frequency. The shift toward red could thus mean only one thing: the galaxies were moving away at velocities proportional to their present distance from us. That is, the farther away the galaxy, the faster it is moving away (Munitz, 1986; Munitz, 1981; Schroeder, 1990). The conclusion, based on this evidence, was clear: our universe is expanding.

Now, with this changing and expanding universe astro-physicists had to propose a theory, to be upheld by evidence, as to how it all happened. Two contending ideas were put forward: one was the steady-state theory proposed by H. Bondi and T. Gold in the 1940s. They maintained that the universe always existed and will always exist, since new galaxies were continually forming, replacing the stars that consume their fuel, cool and die; there was new matter continually created which enabled the universe to renew itself. The theory does not state out of where came the new matter. There is no such evidence available (Hawking, 1989; Schroeder, 1990).

The competing theory was that of the Hot Big Bang, now called the Standard Cosmological Model. It was George Lemaitre who first proposed, in 1927, that "the universe began with an explosion of a highly condensed primeval 'atom' ". His idea was developed and elaborated upon by George Gamow in the 1940s (Munitz, 1986; Domb, 1995). N. Aviezer provides the following brief description of Gamow's theory: "According to the Gamow theory *the universe began 15 billion years ago*. At that time....there suddenly appeared *out of nothing* an enormous source of energy, called the primeval fireball, popularly known as the 'big bang'....The sudden appearance of the primeval fireball marked *the beginning of the universe* in the sense that before the big bang nothing at all existed. Thus, the big bang is the precise realization of creation *ex nihilo*" (Aviezer, 1990).

Despite the difficulty of explaining *scientifically* how such creation out of nothing is possible, in the case of this theory there *is* evidence that such an event took place. The theory predicted that the event of the 'big bang' would still have traces today in the form of weak electromagnetic radiation coming from all over space. Confirmatory evidence of this was hit upon in 1964 by two American scientists, Arno Penzias and Robert Wilson, at the Bell Telephone Research Laboratories. By the 1960s radiation detectors were tremendously improved; Penzias and Wilson "were using an extremely sensitive antenna to measure galactic radio waves. While testing their antenna they observed some very weak unexpected electro-magnetic radiation that seemed to be coming simultaneously from all directions in outer space. It was soon realized that this radiation was precisely what was predicted by the big-bang theory" (Aviezer, 1990). This cosmological discovery of the *isotropic microwave radiation background*, "that is, a radiation background that is constant and equal in all directions of the universe" was of great significance in upholding the theory concerning the origin of the universe (Schroeder, 1990). Not all scientists have accepted the Big Bang Theory. The Swedish scientist Hannes Alfven, winner of the 1970 Nobel Prize for physics, was best known for debunking the Big Bang Theory. Some scientists have pointed out how uncanny a resemblance there is between the Big Bang Theory and the Bible account of the creation, including the commentaries of some of the best known Jewish Talmudic scholars of a thousand years ago! (Schroeder, 1990; Munitz, 1986).

The Universe Observed

Many facets of science have converged and aided cosmological explanations of the structure and origin of the universe. Mathematical physics and observational astronomy and the application of the general relativity theory, have given us a description of the large-scale space-time structure of the universe. The Standard Cosmological Model itself, an evolutionary model which is accepted by most scientists today, has established the idea of an expanding universe: the galaxies are rushing away from each other due to the initial explosion, the distant ones at a faster pace than the nearer ones. On the assumption that all galaxies have the same intrinsic luminosity—the apparent brightness of a galaxy can be taken as a measure of its distance—it is possible to measure the velocity of heavenly bodies through their emission of electromagnetic waves. It is also possible to know the chemical composition of the universe: for instance, the relative amounts of hydrogen and helium that are observed to be present in the universe are in accord with the theory. Each chemical element emits its own unique and complex spectrum of light, which enables it to be identified; as a result, it is possible to assess accurately the abundance of the various chemical elements in the universe. There have also been fresh insights into the way in which various temperatures and energy conditions played their role in the early stages of the development of the universe (Munitz, 1986; Hey and Walters, 1987; Domb, 1995; Aviezer, 1990).

Cosmological scientific knowledge of the kind just mentioned is based on observed evidence, although we must remember that "the observed universe is only a very small fraction of the universe as a whole" (Munitz, 1986). Let us take an important scientific probe like the unmanned US spacecraft Galileo launched in 1989, which entered the planet Jupiter's atmosphere in 1995, sending back, for less than an hour, data on temperature, winds, ammonia clouds and other conditions. The data could prove of great significance for our knowledge of how the solar system was formed, and even provide insights into the original starting period of the universe. Yet in observational terms it explored the tiniest imaginable area, over a minuscule time period, when related to the universe as a whole.

Milton Munitz raises the question as to what are the horizons or limits of the "observable universe." He points out that this depends on such major factors as: "the situation of the observer; available instrumentation; and types of physical theory used in interpreting the observable data, including the conception of the universe as a maximally intelligible system to which the observable universe is related." This means that the horizons of the observable universe are moveable, flexible. However, there are certain unverifiable assumptions and certain facts which put limits on observability. First, the cosmologist is earthbound and confined to our galaxy, although information does reach him from beyond his galaxy. For instance, he learns of the isotropic distribution of galaxies and of the microwave background radiation through direct observational data. Yet the assumption that similar isotropic observations would be made by all other observers at different spatial locations is a hypothesis that is not directly verifiable. To continue with Munitz: "Not only....is the cosmologist rooted in space, he is also rooted in time. Opportunities for observation are severely limited by the duration of an individual person's lifetime, as well as that of the whole scientific community. The time during which such information gathering has taken place or is likely to take place is minuscule compared to the large-scale space-time extent of the universe insofar as this is currently understood. In short, whether taken individually or collectively, the human observer occupies what is for all practical purposes a fixed point in space-time. There are no realistic opportunities for stretching the observer's opportunity for direct observation to match the entire space-time extent of the universe" (Munitz, 1986).

Another aspect of the severe limitations imposed on the cosmologist is that he can have very little knowledge as to where he is located in the universe. He knows that he is in the Solar system and in the Milky Way galaxy, but as to where exactly *this* is located vis-a-vis other galaxies in the universe as a whole, he has very little knowledge—and in certain ways no knowledge at all. Since he can observe the isotropy of the universe he may, at first instance, have some kind of notion or illusion that he is at the center of the universe. But he will quickly give up the idea that our galaxy has any special central position, since by simple logic, "if the universe is isotropic around us it ought to be isotropic about every typical galaxy", which also leads to the idea that "if the universe is isotropic around every part, it is necessarily also homogeneous." These logical arguments mean that the universe seems the same to all typical observers wherever they might be located, and that it looks the same in all directions observed. So, any observer who sees himself, as if, at the center of the world-picture, must realize that any other observer anywhere in the universe could see himself precisely the same way at the center of the world-picture (Trivers, 1985). These hypothetical propositions seem so natural that they have been called "the Cosmological Principle" by the English astro-physicist E.A. Milne (Weinberg, 1983). In view of this principle it would make little sense to ask whether, compared to other galaxies, we are nearer or further from the center or the edge of the universe.

Despite the serious limitations mentioned, scientists using existing cosmic data, observations in particle physics laboratory experiments and applying current theories, have tried to tackle questions concerning the origin and pre-origin of the universe and its eventual fate. The ideas and possibilities are fascinating. The expansion of the universe goes on, apparently for ever. Since there are gravitational forces inherent in the mass and energy of the matter of which the universe is made up, this force prevents it from eventually disappearing into nothing, fizzling out, so to speak. Yet the gravitational forces are not so strong as to make it collapse onto itself, resulting in a 'big crunch', signalling its end, to match the 'big bang' that produced it initially. This suggests that there is a fine gravitational balance at a 'crucial density' of matter and energy. In order to test this last possibility, scientists have had to take into account "the mysterious dark matter which cannot be seen, but nevertheless seems to exert a gravitational influence on stars and galaxies (Gribbin, 1995; Schroeder, 1990). Although taking into account things not actually observed has been justified by most scientists, including the famous physicist Max Planck, who "vigorously asserted the reality of the unseen world" (Perutz, 1989), it is still true that to give satisfactory scientific answers there is need for more accurate and solid evidence. So what we have at the moment concerning basic cosmic questions are scientific ideas which are partly speculative.

Once we pose even more difficult questions we truly enter the realm of speculation. Thus, is the universe spatially and temporally open or closed? Is there something beyond the universe? Has the universe any boundaries and in that sense, can things, disappear off the edge of the universe? Or can the matter of which the universe is composed disappear due to a space-time 'edge'? Is there something beyond the universe, or did anything exist before the universe appeared? Were there any previous events of a universal sort? Was there a vacuum, a pre-existing void, before the beginning of the universe? The questions are overlapping, annoying in a sense, and certainly troublesome for the scientist, whose answers are conjectural at best (Davies, 1977).

Scientists have two options when faced with such questions: They can follow Alan Guth who wishes to escape the bounds of existing theories and provide "speculations on the origin of matter, energy and entropy in the Universe" (Guth, et al., 1983). Of course such scientists will be confronted with certain basic questions, which currently appear intractable, such as the *origin* of mass and gravity. Alternatively they can follow Stephen Hawking, starting scientific explanations with the *singularity* of the big bang, "that crucial interface between total nothingness and the start of our universe" (Schroeder, 1990). Since the existing theory cannot handle questions that go beyond our present understanding of the history of the universe, scientists taking Hawking's option will have to agree with him in saying that "as is the case, we know only what has happened since the big bang, we could not determine what happened beforehand. As far as we are concerned events before the big bang can have no consequences, so they should not form part of a scientific model of the universe. We should therefore cut them out of the model and say that time had a beginning at the big bang" (Hawking, 1989). Stephen Hawking thus sets theoretical limits to the possibilities of scientific understanding of the origin of the universe. As we shall see below Steven Weinberg, a prominent exponent of the big bang theory, points to some limitations which are due to lack of evidence.

Inventing Models

But despite the limited evidence at hand, scientists have proposed explanations of the early development of the universe, its current state, and the prospect ahead. It is interesting to note that before Hubble's evidence of the receding galaxies, and well-before the Penzias-Wilson discovery of the background radiation of the universe, the Russian mathematician Alexandre Friedmann proposed in 1922 two types of model concerning the possible state and fate of the universe: One envisaged the universe as expanding to infinity, should the density of its matter be *below* a certain critical value; the other saw the universe as finite, should its matter have a density *greater* than the critical value, thus producing a gravitational field strong enough to make it implode unto itself (Weinberg, 1983). The first model suggests a universe open and infinite, with galaxies escaping to infinity; while the second model would be a closed and finite universe, eventually falling back into a supremely dense form.

Today cosmologists seem to favor an equilibrium model, such as the Standard Cosmological Model, which we already described above. Based on current knowledge, the following brief history of the universe may be given, so as to arrive at an equilibrium model. Initially the universe was in a state of extraordinarily high temperature and infinite density. Then at the end of the first three minutes, after the big primeval explosion, nucleosynthesis set in. This is the production on a cosmic scale of a mixture of all types of chemical elements from one or two simple types of atomic nuclei, like helium and hydrogen, through a process of large-scale nuclear reactions, of the kind still going on in the sun and other stars. After a relatively long time, once cooling set in and with continuing expansion, the elementary particles were able to clump together into stars and galaxies. The original explosion filled the expanding universe with energy and radiation, with the size of the universe continually changing (expressed in its continually expanding horizon), an unbounded sphere. However, the cooling process seems to have pushed the universe toward a thermal equilibrium; and the gravitational force, emanating from the matter of which the universe is composed, seems to gradually slow down the velocities involved in the expansion, and thus to hold back its expansion. But the gravitational force is too weak to make the universe implode.

Is such a model representing a universe which is truly in equilibrium or just near equilibrium? Is it possible that it is representing a universe that eventually will come to some kind of a halt? Nobody really knows the answers to such questions. No definite predictions can be made, since current evidence is insufficient. Steven Weinberg, for instance, admits that a great deal of uncertainty hangs over the Standard Cosmological Model, and that when it comes to specifics our knowledge is limited. Thus he says: "No one knows precisely what velocity our galaxy has relative to the cosmic distribution of typical galaxies, but *presumably* it moves at a few kilometres per second in some direction" (emphasis added). The understanding of the first few minutes of the development of the universe is essential for the Standard Cosmological Model, yet, as he states, although all the calculations are based on the Cosmological Principle—the assumption that the universe is homogeneous and isotropic—there is no evidence that this principle was valid at the earlier times of the universe's existence (Weinberg, 1983). Weinberg admits that the theories concerning the early universe are highly speculative; that the Standard Cosmological Model cannot be claimed to be true; and that it certainly cannot provide an unequivocal answer regarding the future of the universe (Weinberg, 1983).

Given the uncertainties surrounding the Standard Cosmological Model we may as well indulge in some more fanciful speculation. Howard Trivers refers to one possible cosmic scenario, namely one in which the universe would approach a state of accelerated contraction ending up in infinite density. In such a case, he says: "Some cosmologists believe that a kind of cosmic 'bounce' will take place and the universe will begin again to expand.... If the universe does re-expand, its expansion will again be halted and followed by another contraction, followed by another bounce, and so on for ever....Were this correct, our universe is an oscillating universe, with an endless cycle of expansion and contraction stretching into the infinite past and toward the infinite future. Hence there would have been no beginning and there will not be an end. Thus expansion and contraction constitute the macroscopic rhythm of the universe" (Trivers, 1985). As Steven Weinberg says: "Some cosmologists are philosophically attracted to the oscillating model, especially because, like the steady-state model, it nicely avoids the problem of Genesis" (Weinberg, 1983), and I would add: it also avoids the question of the 'end' of the universe. But as Paul Davies says, the idea of a 'cosmic bounce' and the idea that extinction is avoided by the 'bounce', which occurs because of quantum effects, is purely a conjecture. And "the question naturally arises as to what condition the universe was in *before* the bounce occurred" (Davies, 1977).

It is quite remarkable to compare these 'modern' ideas emanating from the scientific field, with beliefs clearly stated in the classical texts of the *Zohar*, the Kabbalistic or mystical Jewish tradition, and the Talmudic *Midrashim*, commentaries on Judaism's oral law, which have their roots in antiquity. These texts hold that "before God created this world he had created a thousand worlds," and that "God creates universes and destroys them" (*Zohar*, Appendix; *Midrashim, Bereshit Rabba*).

The Cosmos and Human Existence

Some cosmologists wish to go beyond the kind of scientific models we presented and propose that we introduce *the anthropic principle* concerning the universe and the human existence that it contains. This states that the very existence of human life in our universe depends on certain cosmological circumstances; but that equally the very knowledge of the universe depends on the existence of the conscious human observer. To put it another way, although one can assume that there exists a universe that has a set of determinate properties independently of human observation, and that such a universe can exist without human beings, the universe must contain life if it is to be observed and known. As Milton Munitz puts it: "The universe and human life are coupled," and as a result we must move in both directions to understand the relationship—from the universe to man and from man to the universe, "since they are mutually involved in a very special way." This mutuality is stressed by Brandon Carter, who introduced the expression "anthropic principle," meaning that "what we can expect to observe must be restricted to the conditions necessary for our presence as observers." Our human existence "imposes a *selection effect* on what we could expect to observe with respect to the universe" (Munitz, 1986). To to put it another way, "what we see is biased in favour of those parts of the universe where we can exist" (Drees, 1994). This particular relationship between the universe and human life means that the universe, if it is to be cognizable, must have human life; that the presence of human beings 'explains' certain physical constants in the universe, for without those constants life would not exist; and that the constants uncovered by science depend on the selective observation and understanding of human beings. Here it is interesting to refer to certain difficulties encountered in quantum mechanics that were pointed to by A. Einstein, B. Podolsky and N. Rosen in a paper in 1935, which came to be known as the EPR paradox. Briefly stated, they demonstrated that in quantum mechanics the description of reality cannot be complete, so that a probabilistic interpretation is necessary because of unknown hidden variables (van Fraassen, 1991; Sakurai, 1985). We must distinguish between the two theories; succinctly put: the EPR paradox points to the limitations imposed by empirical reality on observability, whilst the anthropic principle highlights the dependence of observable reality on human beings.

But the most interesting upshot of the anthropic principle is that the physical constants of the universe are so finely tuned that any change in one constant would affect the entire structure and ongoing processes, which as a result would mean that life as we know it would not exist. And as Paul Davies points out, "the fact that even slight changes to the way things are might render the universe unobservable is surely a fact of deep significance" (Davies, 1992). To elaborate on this a little, let me quote a prominent physicist, Moshe Kaveh, who in a newspaper article for the general readership puts it as follows:

> Human life on earth depends on the sun; the energy that life on earth receives from the sun is the result of a process of atomic reactions taking place within the sun; these reactions work according to certain physical laws which provide the sun's stability; should these change even just by the tiniest measure, the sun would explode. Again, if the orbit of our planet was, just by a few percent, nearer the sun, or further away from it, human life would not be possible since water, which is vital for life, would either completely evaporate or become deeply frozen; our orbit, one of many possible orbits, is therefore one that is particularly fortuitous and suited for life on earth. (Kaveh, 1994; see also 't Hooft, 1994).

To give another example, in the 1960s Fred Hoyle noted that carbon, which is crucial for terrestrial life, is received from huge stars, which by unique processes produce carbon and release it throughout the universe, yet as F.J. Dyson of the Institute for Advanced Study in Princeton emphasizes, the vastness of the interstellar spaces and consequently our enormous distance from stellar explosions saves us from this deadly cosmic radiation (Aviezer, 1990; Davies, 1992).

So where does the anthropic principle lead us to? It certainly makes us ask a number of questions. Thus: considering that the scientific knowledge we produce concerning the universe depends on existing observations and observability and on existing intuitions and interpretations by certain observers at some given time or times, could there be other observers at some points of time, or the same observers at other points of time, who could come up with different observations, visions, intuitions, interpretations, and so on? Does it all depend, therefore, on the conscious state of the observer, and does this mean that there may be a number of coexisting realities, each 'seen' or 'imaged' by a particular observer in a unique situation? Again, considering that the universe may not give up all its secrets, so to speak, and that, therefore, what *is* knowable may not all come forth in one go, is it possible that the universe we comprehend at any one time is only part of some larger super-universe or a multiplex universe transmitting several messages/signals simultaneously, but picked up differently by various human observers? Finally, could there be observers in the universe, other than human beings on our planet, who 'see' or 'have knowledge' of a universe quite different from that which our scientists comprehend?

In sum, is the universe we think we know just one of a number of different universes possible to comprehend? As Paul Davies tells us, such questions are taken seriously by many physicists and philosophers (Davies, 1992). Thus, according to H. Everett and J.A. Wheeler, the 'relative state' interpretation of quantum theory, which deals with physical phenomena on the micro-scale of electrons and nuclei within atoms, leads to the idea that the possible alternative outcomes of measurement, represent alternative, parallel or successive universes (Shapere, 1984). The same idea was expressed more recently by David Deutsch, a physicist at Oxford University: since in quantum mechanics "objects do not exist in a single state but possess a range of possible states with particular probabilities," this means that "there exist different universes corresponding to each possible state of affairs" (*The Economist*, 1996). Of course, the 'many worlds' interpretations of quantum theory and mechanics, and the very questions posed above, effectively take us out of realistic science and place us unquestionably into the realm of speculation bordering on science fiction.

Testing the Limits of Science

There are many areas studied by science which show the limits reached due to lack of clear evidence or the incompleteness of existing theories. Cosmology is one such area; it has the distinction of showing these limits on a spectacularly grand scale. To exemplify, if an explanation is to be given for the formation of galaxies, inhomogeneities and anisotropy (that is, a lack of uniformity and smoothness), must have developed after the actual creation of the universe, a universe which afterwards underwent homogenization and isotropization. That is to say, without inhomogeneity no gravitational collapse to form galaxies and stars could have occurred. But even if we adduce the quantum theory, according to which fluctuations arise spontaneously, which in the early universe could have given rise to inhomogeneities or clumpiness in the sizes needed to form galaxies, the evidence for this is unobservable, and there is no theory that can explain "why the universe should have any special initial degree of inhomogeneity" (Weinberg, 1983). The problem of the origin and nature of galaxies and the incomprehensibility of creation *ex nihilo* loom large in the uncertainties surrounding the science of cosmology (Shapere, 1984; Trivers, 1985; Parker, 1994).

No doubt, creative imagination and scientific advances, which are unforeseeable, will add to our knowledge in the future (Shapere, 1984). Yet as Steven Weinberg says, ignorance in the area of the early universe is still great, and progress is basically of a conceptual sort (Weinberg, 1983). It seems that the mix of some evidence and a good dose of speculation gives rise to interesting scientific concepts, and may eventually lead to even more fascinating theories. But the theories about the early universe will remain mere conjectures, since no scientist was there to actually see what happened! (Weinberg, 1983).

CHAPTER SIX

IS 'SOCIAL SCIENCE' REALLY SCIENCE?

The root of the distinction between the sciences dealing with the non-human world, known as 'natural science', and 'social science' which studies the behavior of human beings in the context of their social reality, lies in the dual conception of the ultimate nature of universal existence. The dualistic approach sees the all-encompassing universe as consisting of two distinct and independent elements or substances: mind and matter. The same dualistic doctrine sees mind as something unique and entirely different from the materially-based body. It is the mind that distinguishes man from all else in the world.

This leads immediately to the point that where human beings are involved, whether singly or jointly with other human beings, it is their mental states that produce their behavior. They are, therefore, "wholly different from objects in the world of nature." This in turn leads us to suppose that "the areas of the social sciences are different because *persons* enter into them. The natural sciences are concerned with inanimate objects, and even animate objects, but not minds or persons" (Wisdom, 1987). Human beings act with a purpose: they have motives and intentions, and a configuration of thoughts and feelings; these subjective mental states give rise to human behavior. It, thus, follows that in order to study and understand a human act, we must "grasp the meaning with which the actor's intention invested it; a task, as could easily be seen, essentially different form that of natural sciences" (Bauman, 1978; Phillips, 1992).

The distinction we have just drawn raises a number of fundamental questions concerning social science. Furthermore, the time factor poses serious problems for social science, due to the inescapable and intricate links across the time-scale between social behavior and social phenomena (Krausz, 1986). I shall limit myself to only two basic issues: in this introductory section I shall deal with the question whether it is justified for social science to adopt the same rationale as natural science, that is, to make the same kind of assumptions and to use the same methods. In subsequent sections, dealing with particular social science disciplines, I shall focus mainly on the object of study, whether it is the individual and his discernible acts (his behavior); or the larger entities of a social reality, such as groups, institutions, systems, communities, nations and global society; or both in some integrated form. This chapter should provide an overall sense of the *nature* of the social sciences, rather than offer a 'yes' or 'no' answer to the issues raised and to the question in the title. Whether social science is really science ultimately depends on how we define 'science' (see below, p. 127).

Let us then consider the first question posed. If the ultimate ingredients of the human world are radically different from those of the physical and non-human world, then the notions of generalized laws and natural causation may be inapplicable as basic principles in social science. To exemplify: the ultimate ingredients in the human world are thinking persons, whose acts have meanings, persons who have value judgments, persons who may act in a recalcitrant manner, and so on. Not so with atomic particles or micro-biological organisms. On further reflection, however, it appears that micro-organisms do, for instance when disease occurs, behave in what may appear to be a recalcitrant manner, and in their breakdown atomic particles do exhibit random features. Yet any parallels drawn, other than metaphorically, would be far-fetched, since neither physical matter, nor biological living organisms, nor even animals that do have individual feelings, drives and wills, have the kind of consciousness and sophisticated communication that distinguishes the human from the animal world. This highly developed mental capacity, that is called 'mind', is unmatched elsewhere in nature, and it is this that makes human beings into the mind-directed individual persons that they are (*Encyclopaedia Britannica*, 1974).

If we agree that such a radical difference does exist, the conclusion, as we proposed, may be that the basic principles of science are, indeed, not applicable to the human world, since human acts are non-generalizable—they do not follow natural laws, and since causation cannot be clearly established; and that, therefore, there is no point in "imitating slavishly the processes of the natural scientist", empirical processes such as measuring, experimenting, collecting data, statistically analyzing the data, and so on. The same may also be said on the higher level regarding processes of setting up hypotheses, testing them and arriving at explanatory theories with predictive power. The alternative, it has been argued, is to follow the interpretive line of inquiry, which is far better suited for understanding human beings, for uncovering the meanings of their acts, and for making sense of their social world (Wisdom, 1987; Phillips, 1992).

In interpretivist studies, the researcher has the task of intuitively 'understanding' the individual and the meaning of his action or interaction with others, by means of empathy, that is, by placing himself in the situation and probable mind-set of the individual observed. Meanings are often 'objectified' in physical symbols, documents, rituals, and social institutions. These, too, are subjects of interpretive study. This 'hermeneutic' approach builds on *verstehen*—the understanding of the meanings of human acts, that the German sociologist Max Weber (1864–1920) emphasized. It also justifies the classical distinction made in the German language between *naturwissenschaft*, which represents scientific knowledge of the world of nature, and *geisteswissenschaft*, literally knowledge of the spiritual aspects of human life; more generally a sphere of creation and knowledge encompassing the arts and the social sciences.

Those critical of the interpretivist approach argue that the *ad hoc* descriptions, with each social act or phenomenon regarded as unique and requiring its own special interpretation, the results do not add up, and thus cannot lead to general theories of human behavior. Furthermore, it is difficult to know on what grounds the products of interpretive endeavors are warranted. Different interpretations of the same act could be justified. Thus, I may interpret someone's slamming of a door as due to sheer anger, while another observer may interpret it as authoritarian behavior rather than anger. Of course, I may justify my interpretation by the context in which the act occurred. Yet in some cases contextual or circumstantial evidence may help little. In the case of a political 'move' (for instance a politician changing his party affiliation), personal advantage, a change in ideological convictions, and the long-run national interest, could all be adduced as being the *meaning* of that particular act. The looseness of this approach certainly points to a type of study which has no affinity at all with natural science.

This is not only acceptable, it is preferable for those who draw a hard line between natural science and the study of the human world. Yet Max Weber, whose ideas are pivotal for the interpretivists, argued "that a natural science of society was logically possible", but, admittedly, he also believed that it was pointless. It is not generally valid laws that we must seek, as in the natural sciences, but an understanding of concrete reality—concrete causes and effects that are worth knowing. Social reality, including historical reality, does not follow rigid formulae, but rather cultural phenomena which have their uniqueness and which become intelligible "through knowledge of motives, unavailable in the natural sciences". Yet historical and sociological knowledge is just as conceptual and explanatory as knowledge of natural science. For Weber, the radical difference between the two realms of knowledge is obvious (Ross, 1991). In natural science you have rigid universal laws; in social science you discern a general process in particular types of historical situations.

Let us now turn to the views which hold that the social sciences *are* basically akin to the natural sciences. These views are generally called 'positivist' views, and hold that for both the social and the natural worlds only sense experience, and the application of logic to such experience, can provide worthwhile information; positivists reject attempts to gain knowledge through introspection and intuition, the approach used by interpretivists.

Positivism has been criticized by those who believe that the social sciences should not copy the natural sciences. These critics have attacked the naturalistic approach by highlighting extreme positions linked with positivism and labelled as 'empiricism' and 'scientism'. A generally accepted definition of the former would be that any theory we formulate about our world is ultimately dependent on observation and experiment, and that on the basis of empirical observation science can predict and control human behavior. It is not necessary to enter in detail into the dispute between the interpretivist and the naturalist approaches. It will suffice to stress here that from Auguste Comte, who coined the word 'sociology' and who was the originator of the term 'positivism' early on in the 19th century, right up to the Austrian-British philosopher of science Karl Popper, who dealt with these issues in the current century, and including most of those who regard scientific knowledge as a worthwhile aim, the emphasis among naturalists has been on a number of reasonable fundamental requirements of science, including social science. First, we must apprehend the facts of reality—physical, organic, human and social reality. Second, we have to apply logic in order to analyze and understand those factual realities. Third, the ultimate aim of scientific activity is to solve problems (Phillips, 1992).

It must be admitted, however, that the apprehension of facts through direct experience is not free from an interpretive angle. The understanding of what is observed depends on the conceptual baggage, the language and the entire culture of the observer. So does the ability to apply reason in order to analyze the connections between the facts that are gathered. And problem solving is a value-impregnated activity. Scientists will try to solve those problems which appear to them, to their colleagues or to society, to be important; or problems which appeal to them as individuals (sometimes this is because solving a certain problem may become handsomely remunerative). Furthermore, science does include into its factual framework entities, such as certain sub-atomic particles or degrees of intelligence, which are not directly observable or experienceable—which "at best....can be regarded as conventional fictions, as shorthand ways of referring to what *can* be experienced" (Phillips, 1992).

This brings us to a more extreme view in the sciences, one which is diametrically opposed to the interpretivist view, and which is known as 'operationalism'. This approach denies the need to take into account in scientific work the unobservable or unmeasurable entities believed to be present in the world. The operationalist position, championed by the physicist Bridgman and extended to the social sciences by Lundberg, Stevens, and others, is quite explicit on the ultimate goal of science and the criterion on which it is to be judged: its capacity to predict the relationships between carefully defined *observable* measures. This being the case, we only complicate matters by regarding our measures as 'indicators' of something else which is hypothetical and of which we can have no direct knowledge. The social scientist—and indeed any scientist—should concern himself solely with the relationships between variables defined *operationally,* that is, by the observable procedures used in measurement. A variable is, therefore, any measureable characteristic which can assume varying or different values in successive individual cases. Thus, it is alright to study the relationship between sweating rate and scores in an IQ test, but misleading to interpret one's findings in terms of the relationship between anxiety and intelligence—unless of course these words are used in the restricted sense of 'what sweating rate measures' and 'what an IQ test measures'. As Blalock points out, the essence of the operationalist argument is that variables are defined by the operations used in measurement and that without an operational definition a variable has no scientific meaning (Krausz and Miller, 1974; Bridgman, 1928; Lundberg, 1939; Stevens, 1939; Mitchell, 1968; Blalock, 1968).

Although the operationalist position has been useful in stressing that ultimately the test of a scientific hypothesis must depend upon demonstrable relations between operationally defined measures, most social scientists would stress equally the importance of theoretical constructs and intuition, which are guidelines for selecting the phenomena or social facts for measurement and testing. Only by means of both steps can the social scientist provide a meaningful scientific explanation of the social world. Thus, it would be meaningless to provide data concerning the increase in the consumption of doughnuts and the increase in burglaries, since by linking the two no explanation of either phenomenon would make sense; but it would make good sense to link increased alcohol consumption and the increase in the number of road accidents. Most social scientists reject pure operationalism, since intuitive insight into the possible relationship underlying perceived events, actions or behavior is a basic necessity in social science (Krausz and Miller, 1974; Krausz, 1986).

Another approach to the building of social science is that of 'behaviorism' which is related to operationalism, and which had spread especially in American psychology. Behaviorists claim that the task of psychology is not to study the inner state, the 'consciousness' of the human being, since this is not observable. It is rather *the behavior or activities of the human being* that must be studied. Such *behavior* includes all that human beings do and say, and the description of *behavior* must be in terms of *stimulus and response* (Watson, 1925). The best known exponent of behaviorism is B.F. Skinner who seeks to give a causal explanation of behavior patterns of human beings based on science; such explanations do not differ fundamentally from the explanations of the behavior of lower organisms (Skinner, 1953). It is external stimuli which induce animals as well as human beings to behave in a certain manner. Let me quote an example: "by depriving a man of water for a period, or by administering a uretic drug, we can increase the probability that he will drink a glass of water. His action can then be explained without reference to his internal state of thirst, which we cannot directly observe". As Stephen Mennell sums it up: "The causal chain consists of three links: (a) an operation performed on the organism from without (deprivation of water); (b) an inner condition (thirst); and (c) a kind of behavior (drinking). For the behaviorist, any reference to the middle link is redundant..." (Mennell, 1974). As Skinner maintained, in order to be scientific the social sciences must abandon anthropomorphic notions: just as physics stopped personifying things so the behavioral sciences must not appeal to internal states (Skinner, 1971).

This extreme behaviorist approach thus holds that psychology has no concern either with physiological processes of the human organism, such as neural transmissions or chemical reactions in the body, or with mental processes, such as invisible streams of an individual's conscious thoughts and memories. Only overt observable behavior is to be taken into account. Those who do not accept this extreme approach have attempted to find the *units of behavior* which are deemed to be related to certain mental states. To give an example, the turning of a knob with the knowledge that this will bring on light, whatever the reason may be for wanting to have light, relates a unit of behavior to the mental state of knowing what the action does. It is this relationship between behavior and the mind which psychologists with a cognitive perspective wish to examine; however, they would hold that the middle stage—the inner condition of the individual human being—ought not to be eliminated (Seamon and Kenrick, 1992). In the earlier example this was the state of being thirsty, which expresses not simply a physiological need but the mental condition of the individual. Furthermore, the latter varies according to his biography, his current relations to others, his hopes, attitudes, values, and so on. This can be exemplified by the differences in what it means to be thirsty for an individual who is fasting for religious reasons, one who is deprived of drinking because of being a prisoner, or one who is refraining from drinking due to an expected medical check-up. The need to see the more complex aspects of human behavior, which are much more remotely linked to biological necessity has, therefore, been emphasized by psychologists who hold that "behavior has dimensions of organization that cannot be contained within the stimulus-response framework", and by sociologists who stress man's ability to use symbols and language and to attach subjective meanings to their behavior, far beyond anything possible in sub-human animals. It is obvious that social science, which has a human dimension, makes for more complexity, and the question is if *ipso facto* this leads to a less thorough scientific approach than found in the sciences dealing with the physical, organic and animal worlds (Price-Williams, 1964; Mennell, 1974). Let me illustrate the state of affairs in two major disciplines: sociology and economics.

What is Sociology About?

Sociology is one of the least well-understood disciplines in the social sciences, probably because of its very wide scope. It attempts to take into account social reality in its entirety. If a revolution has taken place in a particular country, the sociologist will try to uncover the reasons why it took place, taking into account the historical background of the country, its relationships with other countries, the internal political set-up, the leadership, the ethnic composition of the population, economic conditions and changes, the role of religion in that society, the social divisions, cultural changes and so on. He will try to find the most likely explanations, and he may then compare it with other revolutions, and propose types of revolution which are likely to occur, given certain conditions and factors at work.

So, what is really the subject matter of sociology? One could give a very broad outline by saying that basically sociology aims to analyze and understand the structure and the operations of society, including all the social phenomena that occur in it. To achieve this, the sociologist must study the patterns of social relationships and the framework within which these unfold. The framework is a given society which may or may not be coterminous with national boundaries. One may look at contemporary England, feudal Europe, or a much smaller local community, but the aim would still be essentially the same: that is, to discern the functioning of these entities in terms of some of the major social institutions. *Social institutions* are the established arrangements and ways of doing things in different areas of society, such as the system of law and government, economic arrangements, family life, religion, the educational set-up, recreational arrangements, and so on. In this sense even war may be regarded as a 'social institution', since it is one of the common ways in which human beings carry out their disputes. Thus, the sociologist would say that war has been *institutionalized* in human societies.

There are also other ways of looking at sociology. Some sociologists stress that their subject deals with the organization of human beings into groups. According to this view the principal aim of sociology is to discover how groups operate, how they relate to one another and how they influence the lives of their individual members; how they change and why, and why some grow and flourish while others decline. The largest and most inclusive of the groups that men have formed is *human society*, a group in which the members share an overall common way of life. Society, therefore, constitutes the framework within which most of other human groups function. Just as it is impossible to obtain an adequate understanding of specific human organs, such as the heart or lungs, without reference to the larger biological system of which they are part, that is the human body, so, too, it is impossible to understand the smaller and more specific human groups, such as individual families or communities, without reference to the larger system of which they are part, that is human society (Freedman, 1956).

There are those who view sociology as the discipline that studies how people interact and the effects of interaction on human behavior. There are different kinds of interaction, ranging from the physical contact of the newborn baby with its mother to a philosophical discussion at an international conference; from an informal greeting in the street to the formal meeting of a nation's cabinet; and from the cooperative interaction of soldiers and officers in the army to the conflict between a rioting crowd and the police. What processes lead to the interaction, what exactly occurs when they take place, and what are the short-run and long-run consequences of them are the subjects sociologists study.

It is true that not all human behavior is controlled by these interactions because there are other factors, such as certain biological and psychological aspects of human behavior which are independent of such interactions. However, man is distinctive among animals in having a special means of interaction—namely language or the communication of significant symbols; and the behavior of man, as distinct from the behavior of other animals, is predominantly influenced by these interactions. Man becomes human in his behavior through interaction with his fellow human beings; thus sociology is *par excellence* the study of what is distinctively human about man. But it is not just the interaction that takes place at a particular time that is important to the sociologist, since the individuals involved and their thoughts are based upon and influenced by the accumulated thoughts and ways of behaving of their whole society going back into history; and similarly their behavior will influence others in the future. It is the accumulated thoughts, behavior patterns and their consequences that constitute *culture*, and culture provides the overall common understanding that makes society a viable social system (Rose and Rose, 1965).

The above ways of looking at life in society, that is the *institutional aspect*, the *group aspect*, and the *interaction aspect*, have a common thread between them. They all imply that there is some kind of orderliness in social life, and that, therefore, this lends itself to systematic scientific treatment. It is as well to put forward, therefore, the definition of sociology by Imogen Seger: "Sociology is the systematic and controlled observation and explanation of regularly occurring social relations, their causes, conditions and consequences" (Seger, 1972). This is not to say that sociologists do not recognize the existence of disorder and change. They also have to contend with the difficulty of taking into account certain seemingly unpatterned aspects of human life. There are, for example, the unpredictable chances of war, the ambition and genius of an extraordinary individual, or the effects of a fortuitous invention of some new technique. The answer may be that as much as such events or individuals jolt the system, and important as some changes, such as industrialization, may be, society will continue to be based on the interdependence and interaction of individuals within groups, and in all cases individuals will be influenced by the social institutions and the culture of their societies; bearing in mind, though, that institutions and culture are subject to dynamic processes.

In all that I have said so far about sociology it is obvious that the focus has been away from the individual. It was Emile Durkheim (1858–1917) the French sociologist—one of the classical thinkers and founders of the discipline—who emphasized the external reality of society and the need to study *social facts*, which are external to the individual and which control his conduct. One of the main examples which Durkheim used to illustrate what he meant by 'social facts' is law. One of the features of law is that it has an existence separable from individual acts of law enforcement, so that it can be studied apart from the individual. In some cases, say in a primitive tribe, where the separate embodiment of law was not obvious, it was still the case, Durkheim argued, that the individual's conduct was influenced by the "collective current" existing in society. This influence can be recorded and studied by means of statistical data concerning the behavior of entire populations. Thus, statistics can help us ascertain the social causes of delinquency or suicide by discovering the different frequencies of such occurrences in particular populations with specific characteristics and in specific situations (Durkheim, 1938; Durkheim, 1956).

Many modern sociologists, likewise, see the principal task of their discipline and of the social sciences as being that of explaining social phenomena, and not the behavior of single individuals (except in some branches of psychology). For James Coleman the ultimate aim in studying social phenomena is to understand the behavior of social systems. In this endeavor he sees as legitimate to study the behavior of social systems either by the observation of a number of systems *in toto* over a period of time, or by examining the processes internal to the system; that is, by moving to a lower-level of the component parts of the system, such as institutions or individuals (Coleman, 1990). It may sound strange to talk about the *behavior* of systems; but for Coleman actors are not necessarily always individuals. In modern society, especially, there is an abundance of corporate actors such as government bureaus or trade unions—the corporate actor being composed of *positions* occupied by persons. It is corporate action that characterizes the constructed social environment of modern society. The actor, whether individual or corporate, carries out the social exchange within the *system of exchange*, such as that of a competitive labor market. This means that individuals are linked to collectivities, something that we see even more vividly in such collective behavior as riots, panics, or fashions (Coleman, 1990).

It is rather surprising that a great deal of debate has been going on in sociology around the question as to whether entities described as 'social wholes', such as the State, the legal system, or institutions such as families or a financial transaction, actually exist apart from the individuals who compose such systems and institutions; or whether only the individuals actually exist—and they, by their social acts give rise to holistic entities 'seen' by social scientists. Thus, social scientists construct the social reality they examine. Is this constructed social reality of the social sciences an imaginary world? To answer such questions sociologists have pointed to social phenomena which give expression to the existence of 'social wholes' as true realities. Examples may be the marriage ceremony and the physical location of the married couple and their children in a certain type of household (which signify the existence of the family), or the gathering of representatives of a certain population in a Parliament and the declaration of the Prime Minister (acts which express the existence of a State). Sociologists go even further: while recognizing the intentional and meaningful acts of individual human beings, they point to the social character of language and of meaning to stress that every human being is society-dependent, and highlight the fact that without the process of socialization we could not even conceive meaningful human behavior. Individuals, after all, are born into society!

The 'social facts' or 'social phenomena' of the sociologist incorporate a very wide spectrum—much that is not directly observable, much that is implied, much that depends on both social convention and individual attitudes and interpretations. Social phenomena are the outcome of a complex interplay between physical environments, bodily needs, individual actions and wills, interaction between individuals, social institutions and symbols, group organization, culture and history, and also visions or expectations of the future. In real life such social phenomena surface in the form of, parliamentary legislation, a strike, a political view expressed by an individual, a dispute between an employer and an employee, a game of chess, a love letter, the move from one occupation to another, a religious ceremony, and so on. But these occurrences are not one-time occurrences, exclusive to some one individual only, and although each occurrence has its peculiarity, specificity, or vantage point, such occurrences have common characteristics which allow categorization and quantification. In this way the sociologist produces aggregated data: for example, the number and type of contraventions of certain laws passed by the parliament, the frequency of attendance at religious ceremonies by different age-groups, or the number of divorces in a given population. Furthermore, certain connections between such occurrences, or social phenomena as we call them, may be postulated, such as between the frequency of strikes and the percentages of job dismissals in a working population. In scientific language these postulations will be the 'causal hypotheses', the corroboration of which is sought in the aggregated statistical data. Such data can be gathered and aggregated not only for social phenomena which have an 'appearance' and are, therefore, directly observable, but also for motives, attitudes and feelings which are expressed by individuals, or found by the social scientist who implies these 'states' from patterns of behavior.

Assuming that the arguments so far are sufficiently convincing for us to accept that the sociologist *can* apprehend social phenomena, either by direct observation or by indirect evidence, even then he still faces four major problems: (a) Does the complexity of the social world prevent him from using the natural science causal model? (b) Does his inability to test experimentally his causal postulations deny him any truly scientific conclusions? (c) How does he overcome the difficulty of working at two different levels: data collection at the micro/individual level, which he transforms into aggregate data for analysis at the macro/group or system level? (d) How can he reconcile the apparent contradiction between the existence of choice of action, due to options available for the individual actor, with the existence of institutional and societal necessities and constraints? We shall now proceed to examine each one of these questions.

Let us take now the first major problem, that of complexity. Stephen Turner demonstrates this by saying that not only are there many possible causal factors to an individual human action, but also that the causal process leading to the action is routed through the cognitive processes going on in the individual himself. Supposing we take the delinquent act of a juvenile: here we must consider a whole host of possible causes, such as violence in television, the decline of the West, his genetic inheritance, the neighborhood, a demeaning encounter with an authority figure, the opportunity available to carry out the act, and other possible factors. Since the possible influences of such causes are played out in the cognitive mental processes going on in the individual, it is impossible, argues Turner, to disentangle this complexity and arrive at a clear cause-effect sequence. The core of the problem is the human action itself (Turner, 1982). One can counter this point by arguing that the very strength of the use of statistics to cover a small group of people or a larger population (that is, the aggregate-holistic approach), is that it offers the possibility of avoiding reference to individual processes or particular events routed through the mental realm. And this avoidance is justified, since the sociologist seeks to uncover the frequencies of certain acts or attributes of human beings in certain classes of people or in some population, acts or attributes which are deemed to be relevant to the problem or topic under scrutiny. Thus, we are entitled to assume that the sociologist is not interested in the social mobility chances of any *one* individual, but rather in the mobility trends that obtain in certain populations in particular periods. Similarly, the sociologist is concerned with trends in delinquency rather than with the causes of any *one act* of delinquency.

However, it is true that even when we consider the substantive field from this aggregate point of view, we discover elaborate patterns of highly correlated events. To add to the difficulties of the sociologist, these complex patterns cannot be sorted out by experimentation, which is the second major problem posed. Not only are there economic and humanitarian factors which prevent experimentation, but the experimental situation can never truly replicate the natural social situation where social acts are carried out and where phenomena occur, and the experiment itself could have a confounding influence. The sociologist usually resorts to the possibility of carrying out quasi-experimental control. Thus, he could take two randomly chosen populations, giving for instance one of the chosen group instructions and information concerning the use of water in the summer, but not providing such instructions and information to the other group, and subsequently observing any differences in the average level of water consumption in the two groups. The effect of the instructions/information could thus be gauged. Another method resorted to by sociologists is to set up models for the analysis of the great number of variables usually at play. Such models must take into account the structural and contextual effects on individual outcomes, as well as the cumulative effects of individual attributes and behavior.

Thus, the performance of a student in a knowledge test will bring into play such variables as his innate intelligence, family conditions, his educational history, his beliefs, attitudes and values, as well as the formal structure of the school and the social context of his school and out-of-school activities. Such models usually show the direct effects of some of the variables as well as the indirect effects by cross-links between variables; say, innate intelligence will have a direct effect on the student's performance, while family conditions may affect it via the sort of school he attended and the attitudes he picked up in the family. These effects are usually shown by calculating statistically the strength of the correlation between the influencing factors (*independent variables*) and the outcome (*dependent variable*). The problem faced by sociologists is that often these effects, which explain the outcome, do so only partly: often they do not add up to more than about fifty percent of all the possible influencing factors, the rest being residual unknown factors. The sociologist takes refuge, in this instance, in the uncertainties surrounding many natural science fields. The best example is medicine, where often the etiology of a disease is not fully grasped, and medical science can point only to some of the possible causes of the disease. Time and again, science has to make do with partial knowledge only.

Let us now look at the third problem mentioned: the different levels at which data collection and analysis are carried out. Data collection is always at the micro-level: the sociologist gathers information concerning the behavior of the individual, say his decision to take up a certain occupation; the interaction between individuals, say the playing of cards; and some larger group events, such as a political demonstration. In all these cases he is interested in gathering information from many individuals, he wishes to observe many small interactions and as many larger occurrences, such as demonstrations, as possible. Even where his aim is to understand how parts of a system, such as departments in an organization, combine to produce certain results, he will unavoidably refer to a number of typical combinations and results in a number of organizations. For the analysis of aggregated data the sociologist must classify the events he observes, which is to say that he will establish classes or categories or types of social behavior and social events. It is in this way that he is able to uncover links between social phenomena, and trends and processes in social structures. A typical example would be the link between changes in family structure from the extended family to the nuclear family and to the single-parent family, and the development of modern industrial society. I shall not go into the methodological issues of how the sociologist constructs his categories and why he must resort to flexibility in his conceptualization of categories in order to make his results highly relevant to the subject he studies, while preserving the possibility of comparability by retaining existing categories used in other researches on the same topic. There are ways of achieving both—by producing sub-categories within larger categories, or by 'collapsing' smaller categories into larger ones. To return to the question posed, there is no problem in gathering data at the micro-level and then carrying out a macro-analysis. But it will become clear, when we shall deal with the final problem of individual intentions and choice, that the sociologist cannot move back from his group-focused analysis to the causal explanation of any *one* individual's behavior.

In order to deal with the last problem, the individual's ability to choose between his actions despite the social boundaries and constraints within which he can act, it is worth pointing out with Daniel Dennett that there are "societal pivot points", so that people's choices depend on a particular belief X or some desire Y, and not on any of the other infinitely many ways in which they may be alike or different from other individuals (Dennett, 1981). I am inclined to argue further that since most of a person's acts are frequently repeated by that person, and since we constantly find similar, indeed almost identical, acts carried out by other individuals too, the same societal pivot points are used, given particular conditions or circumstances, so that the intentional behavior of an individual may be considered *not unique*. In other words, the independent decision-making of the individual is always affected by the culture, economy, polity, etc., in which he exists. The decisions of the individual are always bounded by the norms and constraints of the system. In fact he is aware of the social system and he refers to it in his decisions, even though he is often unable to grasp it fully. Therefore, notwithstanding his intentions, the encounter between the individual and the system often produces unintended consequences. Both because of the unintended consequences and because the individual is not usually in a situation of totally free choice, but neither in a situation of forced choice since he has options within certain limits, human behavior is not entirely predictable (Boudon, 1981).

To sum up, it is clear that sociology, in dealing with human behavior in society, can unify or at least bridge the dual nature of this human reality: the individual as a thinking, communicating and decision-making entity on the one hand, and on the other hand the social structures which this reality gives rise to, encompassing the persistence of certain types of human groups, the culturally patterned practices, expectations and norms, and the established and institutionalized relationships between the interacting human individuals. We can see this unification or bridging on the empirical as well as the theoretical level of sociology.

On the empirical level the sociologist gathers data about many individuals, data about numerous interactions, about patterns of social life in many groups, and about a multitude of events taking place in human societies. He lumps his data together according to certain criteria in order to carry out his scientific analysis. How is the sociologist's 'science' affected by the fact that, at the base, the data he gathered and arranged into categories are composed of single intentional acts of individuals? I wish to put forward the view that such intentional behavior does not destroy a social science based on aggregated data. It does produce, however, a degree of uncertainty, since the 'choice' factor of the individual acting intentionally becomes the 'chance' factor in the statistics which result from the gathered data (Krausz, 1992). The same applies to decisions taken in groups and on behalf of groups, or to the outcomes of, say, informal interactions between people. In all such cases social statistics will provide only patterns and probabilities. Let us take the simple example of an individual choosing, for whatever reasons, to travel on a suburban train on a particular day. On the same day there will be many thousands of individuals choosing to travel, for their personal reasons. But the sociologist does not have to probe each individual decision-making so as to provide a sensible estimate of the number of passengers on that day. Knowing the demographic factors (the number and age composition of those living in the suburb), the number of car-owners, parking facilities, and the main reasons—such as work and shopping—that induce people to travel to the city, will suffice to make the estimate. The social scientist's observation and data gathering will concentrate on the sort of information that will enable him to examine and explain the probability of certain social patterns of behavior taking place. Individualized behavior, and the social phenomena this gives rise to, will fit, therefore, into the explanatory scheme of macro-quantitative sociology. The statistical rules of large numbers are applicable.

On the theoretical level there have been numerous propositions of how to link, and thus understand, the relations between the individual and society. One such theory is that of Anthony Giddens who points to the individual as a creative agent and a partner with society in producing social structures. Individuals are "meaning makers": *they* give the meanings to their acts. As such, they not only act according to social rules, but as reasoning agents they have the ability to change and create those rules and the meanings attached to them. Therefore, Giddens sees a continuous process of "structuration"; that is, social relations *are* structured, but they are continually changing. Social structures are not *automatically* reproduced; their reproduction is in the innovative hands of human agents (Giddens, 1984). In this way there is a historical dynamic aspect to social life: the social rules created in the past influence the present, and the changes taking place now will have their effects on the future. This does not necessarily imply an evolutionist development; outcomes of the continuous process of structuration may be quite unexpected, and there are no clear indications of a particular linear manner of structural changes towards, say, the 'good society' (see for example the 'good society' described by J.K. Galbraith, 1994).

Such a theoretical view leads us to conceive society as being in constant flux, but never without the structured relationships of its component individuals and groups, providing sufficiently stable expectations and thus making life in society manageable. As Anthony Giddens says, neither the human agent, nor society, nor social institutions should be regarded as having primacy. The notion of human "action" presupposes that of "institution" and vice versa (Tulea and Krausz, 1993). This mutual relationship is the core of the account of how it is that the production and reproduction of social practices, across time and space, takes place. Giddens, of course, provides a more detailed and sophisticated account of this process of "structuration" (Giddens, 1982). The contribution of Giddens, and other theoretical sociologists, lies mainly in providing us with a descriptive meta-theoretical framework, that is, the establishing of the principles upon which human existence in society are imagined to be based.

Economics as Science

Our second example of a social science field is the discipline of economics. Its modern beginnings are to be found in the eighteenth century writings on what was known as political economy. The most famous writer on the subject during that century was the Scottish political economist and social philosopher Adam Smith (1723–1790), whose major work *An Inquiry into the Nature and Causes of the Wealth of Nations* (1776) has had a long-lasting effect. His study was concerned with the accumulation of material wealth in the conditions of the final advanced historical stage toward which human society was supposed to evolve. His contention was that, if not impeded, this stage would bring forth a laissez-faire economy ruled by the market-forces of free competition and characterized by a sophisticated division of labor; these developments being conducive to an increase in productivity. In more modern parlance such a *capitalist market economy* would enhance the wealth of the nation where these conditions prevailed. For the emergence of political economy as a science, Smith's view of humanity's historical development was far less important than his exposition of the invisible hand of competition which was guiding the economic system based on individual self-interest. Once it was accepted that such a self-regulating system existed, the possibility of a new science of economics was recognized, and the subject matter of this new science became the production, accumulation and distribution of material wealth governed by inexorable 'laws' (Hausman, 1994; Kirzner, 1976; Wicksteed, 1933).

During the eighteenth century the English political economists Jeremy Bentham (1748–1832) and Thomas Malthus (1766–1834) shifted the focus from wealth, by showing concern for the inescapable poverty of nations. Their economic pessimism together with Karl Marx's emphasis on humanity's constant struggle for the downright necessities meant that on the whole the materialistic outlook on economic affairs became even more pronounced. Economics acquired the label of the "science of subsistence". In the view of some economists, economics was narrowed down to "the study of the goods necessary for human survival" (Kirzner, 1976).

Another basic principle enunciated in classical economics in the 18th century, stated by Adam Smith and more rigorously restated by David Ricardo (1772–1823), is that of 'comparative advantage': it is not only beneficial for a country to specialize in those goods which it can produce at a lower cost than other countries, but it can benefit even more if it specializes in goods where it has the greatest comparative cost advantage. To apply this principle it was necessary to support free trade between countries, which would lead to an international division of labor. As David Hume (1711–1776) emphasized, free trade increased real income by bringing about a more efficient allocation of resources (Blaug, 1968; Rotwein, 1968).

Some of the broader definitions of classical economists, seeing economics as being concerned not just with wealth but also with human welfare, did not change the economic point of view. Since welfare was conceived as *material* welfare, economics continued to be seen as a science dealing with a particular sector of reality, that is the material conditions of human existence. An important change in the definition of economics occurred only when a number of economists shifted their attention to the individual's behavior concerning his material subsistence. Now questions were asked about how he is motivated. The seeds were there in the views of Adam Smith and David Ricardo, that it is self-interest that regulates the economic system. Economists began to adopt the assumption of a *homo oeconomicus*, namely that, so far as human activity connected with material existence is concerned, the individual's motivations stem from his 'selfish nature', which can be expressed in the less value-loaded term of 'economic man'.

As Kirzner points out, the break was now clear: "Once economics was conceived as involving a certain pattern of *behavior*, or even a uniquely motivated kind of behavior, then the bonds that attached to the class of objects constituting wealth could easily be broken....Economics could then be identified, not in terms of wealth, nor even in terms of men-in-quest-of-wealth, but in terms of a unique pattern of human behavior: *the getting of the most for the least*. This pattern of behavior came to be variously known as conforming to the 'economic principle', as obeying the 'law of least means', the 'maximization principle' and the like" (Kirzner, 1976).

The economic man trying to satisfy his wants will be guided by the economic principle when he engages in exchange activities with other individuals in the 'market', namely the economic system. He will try to economize the means whereby he can achieve his ends. Of course, all the individuals in the economic system are 'economic men' and will act accordingly. In other words, they will all try to strike the best bargains vis-a-vis the others. This is because the economic man realizes that his wants are many, but that the means at his disposal are limited. The idea of scarce means was long recognized, for instance by Thomas Malthus and by the physiocrats. The latter were a group of French economists, in the eighteenth century, led by Francois

Quesnay, who believed in the operation of natural economic laws and championed free trade, and who excluded 'free goods' such as air, from the subject matter of economics, since being abundant they were not objects of exchange.

If we pause for a moment and sum up the ideas that developed in classical economics, from the eighteenth century right up to the beginning of the twentieth century, we may get a feeling that economists have endowed man with what might appear to be a not very praiseworthy character, which made possible, however, the formulation of the principles of their science. Thus, man was seen in his behavior as consciously a cold-calculating person, who seeks to maximize his satisfaction with the least cost to himself, which could also imply that he achieves this at the expense of others; a person never fully satisfied and constantly driven to improve his own position. This concept of a *self-centered economic man*, however, set up the basic principles for the science of economics: man must be a calculating person since he always faces choices, hence he must arrange his wants in an order of preference and calculate the relative costs involved in achieving them, and thus decide according to his preferences and the costs, his course of action. By the 1930s these ideas became crystallized in the best known definition of economics by Lionel Robbins, who laid down that: "Economics is the science which studies human behavior as a relationship between ends and scarce means which have alternative uses" (Robbins, 1935). There are a number of core principles that arise from this definition. (a) Economics is concerned with human activity which is affected by the prevailing scarcity of the means available to satisfy ends; (b) Human beings cope with the situation by means of rational thinking when making their choices; (c) Economizing and making choices applies widely, including for instance the use of time; it affects not just individuals but also groups of people such as families, and entire nations; and the interacting choices give rise to certain forces or mechanisms which ultimately determine the outcomes.

Let us look at these principles in somewhat more detail. First, it is the scarcity situation which produces the economic problem. But does this always apply? What about the multi-millionaire who has so much that he cannot be thought of as facing an economic problem? C.D. Harbury makes the point, however, that even such a millionaire is faced with the economic problem of how best to spend the twenty-four hours of the day. Similarly, affluent societies also face the choice of how to apportion their wealth: the more is allocated to private consumption, such as three cars per family, the less will be available for public goods, such as roads or space research. The old adage 'guns or butter' is still true, even for affluent societies today. No wonder that economics "has been dubbed as 'the science of choice' " (Harbury, 1971). Arising out of this situation of scarcity and the need to make choices we get to a central concept in economics: *opportunity cost.* This can be well exemplified in the field of production of goods. In order to produce goods, labor and capital are necessary. But these factors of production are scarce, so that a choice to produce armaments means that the chance of producing food has been sacrificed; that is, the opportunity for producing food has been lost, and, therefore, the opportunity cost of the armaments is the food that could have been produced had that been the choice (Harbury, 1971).

The second core principle is the use of rationality in choosing between alternative and differently valued ends. Economists elaborating on this principle make a number of assumptions in order to arrive at a general 'rational choice theory'. Some of these assumptions are quite basic, such as that the individual knows what he wants, that he is able to order his preferences, that as a consumer he will shop around to ascertain where he can purchase goods of the quality he wants at the lowest prices, that on the whole he will try to maximize his satisfaction; or as economists put it: people seek to "maximize their utility". Now it must be recognized that there are difficulties in fixing a paradigm (that is, the central and overall way of regarding phenomena within which a scientist carries out his work) based on rational behavior and rational choice. The difficulties are: (a) that the individual's choice may turn out to be not to his benefit or may not be according to his expectations; (b) that an individual may act in an altruistic way; (c) that the information on which he bases his choice may be incomplete or faulty; (d) that individuals often act impulsively or out of habit. There are several possible answers in defense of the rational choice paradigm. First, that due to the imperfections of reality, such as the lack of adequate information, the individual whose choice *is* rational is apt to make mistakes, and therefore it could be argued that rationality exists but is bounded by reality's constraints. Second, that the existence of irrational behavior is not denied, but that the economist looks at that part of reality where rational choice is paramount in determining the individual's action (McKenzie and Tullock, 1978).

Now let us take the third core principle, namely, that the choices of interacting individuals, or interactions between groups organized into economic institutions (such as firms, sectors of the national economy, entire national economies) will lead to certain outcomes, to be seen as levels of satisfaction or efficiency. These outcomes will depend on the existing conditions, namely the resources of the individual, of the firm, or of the national economy, and the constraints of the reality within which the interaction takes place. Once the interactions take place a mechanism will be at work which determines the outcome. The simplest example is the price that individuals will have to pay for a certain product which in a free and perfectly competitive market will be determined by the forces of demand and supply. The price paid will reflect the maximum level of satisfaction achievable or the greatest efficiency in allocating resources, given present conditions. We may add here that the individual's demand and the price he will be willing to pay will depend on the utility or satisfaction he derives from the goods in question. This is subject to the law of *diminishing marginal utility*. Put simply, the first bar of chocolate will give me great pleasure, but by the third bar the amount of pleasure I may get will diminish considerably. Therefore, the more chocolate I am being offered the less I will be willing to pay for it. The greater the supply, given a level of demand, based on perceived utilities by individuals, the lower the price, and vice versa, the more limited the supply, against the same demand, the higher the price.

Without entering into the finer points of price theory, two aspects are worth mentioning. First, that certain changes may operate to bring about a different situation in the market: thus, tastes or fashions may change the demand on behalf of consumers, or a new technology may produce a flexible supply. Economists are fond of using the caveat of *ceteris paribus*, that is protecting themselves in the manner of 'other things being equal': saying that if conditions do not change the forces at work will result in an equilibrium, represented, say, by the price of coffee, the share-index on the stock exchange, or the exchange rate of two currencies. Should conditions change, however, a new price level will emerge. The second aspect I wish to mention is that considering the ability of individuals, or firms for that matter, to order their preferences concerning the many goods they wish to acquire, they can also order the different combination of goods they wish to have, given the different prices of commodities in their combinations. An individual may find that in a *scale of preferences* he is indifferent to two combinations: that is, he is equally satisfied with combination A (a holiday abroad, plus new furniture for his living room, plus private schooling for his children) and combination B (a more expensive car plus private schooling for his children). One thing is clear: he will endeavor to stay at a level which allows the combinations that he equally desires (Stonier and Hague, 1953).

Again, according to the science of economics, what actually happens is that the individual will arrive at a set of combinations which he knows he cannot improve upon, without causing others to worsen their situation. This is a state of equilibrium known in economics as the 'Pareto optimum'. Vilfredo Pareto (1848–1923), a Franco-Italian sociologist, developed the theory of maximum efficiency, known among English-speaking economists as the concept of 'optimum resource allocation'. This is where it is impossible to increase the efficiency of one's resource allocation and thus increase one's satisfaction, without decreasing the satisfaction of others at the existing levels of available resources. Of course, should the overall availability of resources in the economy change, the optimal situations for all individuals will change too (Allais, 1968).

Although one can extend the theories which built up the science of economics from the behavior of the individual to the firm, to economic institutions such as banks, and to the whole national economy, it is true to say that classical economics has been concerned with economic forces at the micro-level, that is, its aim appeared to have been that of taking a close look at each part of the economy, rather than at the interrelationship of the parts, and basically of analyzing resource allocation through the price mechanism. It is also true that the assumptions underlying these theories are the kinds that apply to 'pure' situations such as 'perfect competition in the market'.

The idea of the price mechanism in the micromarket situation stems from the 'partial equilibrium analysis' of Alfred Marshall (1842–1924), a prominent British economist. His approach highlighted the particular equilibrium attained in a micromarket, given that static conditions prevail on both the demand and supply sides, so that at the equilibrium point demand equals supply. According to this principle if one of the assumed (*ceteris paribus*) conditions changes, say consumer tastes or import duty, a new partial equilibrium sets in. Marshall made it clear, however, that his "ultimate objective was to develop a full-fledged theory of dynamic change and growth" (Corry, 1968). But Marshall realized that we cannot assume every action to be deliberate and the outcome of calculation, and defined economics in wider terms as "the study of men as they live and move and think in the everyday business of life; the economics that temper the objective logic of choice with the subjective traits of ordinary people" (Maital, 1982). Despite this broad outlook, and despite a deep concern about poverty and, above all, his belief that human beings are capable of change and progress, Marshall came in for a great deal of criticism for diverting attention from the real macro-level issues. The criticism mounted against Marshall in particular—and classical economics in general—during the Great Depression of the 1930s, for ignoring the true face of economic reality.

Many of the forces at work were not taken into account: the power of monopolies in fixing prices; government policy so far as the level of employment was concerned; unexplainable ups and down in the economy, labelled as 'business cycles'. Economics entered a period of crisis especially when John Maynard Keynes published, in 1936, his book *The General Theory of Employment, Interest and Money* which "precipitated a revolution in economic theory" (Hausman, 1994).

The arguments Keynes put forward were as follows: A prolonged depression with long periods of high levels of unemployment was possible. This is something that classical economists denied, since they relied on the self-correcting mechanism of the market. However, the latter did not work once people lost confidence in the economy and started hoarding money. The resulting lack of demand would not necessarily bring about a drop in prices, since the price of labor, that is wages, as a result of the intervention of trade unions, would not easily decline. Even if wages did go down, this would further depress demand, and thus discourage investment in industry and the economy as a whole. The whole cycle of exchange would slow down and the economy would settle, for a period, into an underemployment equilibrium (Schumpeter, 1994). Keynes further argued that the situation could be changed by government intervention; thus, with increased government spending, the aggregate demand for commodities could spur the economy out of its stagnation. Fiscal and monetary policy was, therefore, of utmost importance: increased government spending, say in public works, lowering interest rates and providing easier credit to industry and to individuals, could result in increased productivity if a sufficient part of this monetary liquidity finds its way into industry and increases its efficiency. So, much depends on the propensity to consume. Should the latter be very high it could push demand beyond the productive capacity of the economy, if the latter has already reached a situation of fully employed resources. This could start a rise in prices due to a pressure on the inadequate supply, which raises the costs of resources when such a situation is reached. "Costs, prices and wages tend to chase each other", resulting in an inflationary spiral, and could even lead to a 'runaway' or 'hyper' inflation, and result in a total loss of public confidence in the currency (Harbury, 1971; Wallich, 1994). Again, Keynesian theory would rely primarily on government controls in order to rectify such a situation.

It is beyond this brief review to consider the views of economists concerning the causes of business cycles or the ways in which economic booms and periods of stagnation or downturns are started. What is important to emphasize here is how Keynesian economics affected the discipline of economics. Two main points arise: 1) the development of macroeconomics and applied economics; 2) the questions raised concerning classical economics, and the controversy over a definition of economics based on the assumption of scarcity.

Regarding the first point, it is clear that the emphasis on understanding how a particular economy works in its entirety, and the application of microeconomic theories and models to macroeconomic analysis, has taken center stage during the second half of the twentieth century. Robert Solow, the 1987 recipient of the Nobel Prize in economics, expresses this by saying: "Macroeconomics is what it is all about. If you do not understand the business cycle, unemployment, inflation, the real exchange rate, well you do not understand economics at all". In his view, attention must be paid to the "macro-foundations of microeconomics", in that "we are entitled to ask what sorts of microeconomic mechanisms both look right and do justice to the nature of the general economic environment in which they are expected to function" (Sichel, 1989). The second point of emphasis emerging from the Keynesian approach is that classical economics was wrong in assuming that "the idleness of resources could be only a temporary phenomenon of disequilibrium". An underemployment equilibrium appeared as a fact (Kirzner, 1960). It follows, therefore, that the absence of scarcity of resources, in the words of Barbara Wooton, "renders inoperative, irrelevant and unreal the *whole* corpus of economic studies as defined by....Robbins and as embodied in the classical analysis and its contemporary elaborations and refinements" (Kirzner, 1960).

So the Great Depression of the 1930s inspired Keynes to launch his macroeconomic thinking and to ridicule the classical notion of a self-regulating economy. The new theory, as we indicated above, advocated active government fiscal and monetary policies, such as changing the government budget or varying the money supply, in order to manipulate aggregate demand by households, businesses and government departments, and thus achieve full employment and price stability. It was also thought that government intervention, by means of taxation, subsidies and trade regulation, could correct the inefficiencies. The argument was that the orthodox competitive-model did not work in reality and that, therefore, government intervention was necessary. Around the issue of government intervention there evolved the new school of 'public-choice economics', defined by Dennis Mueller as "the economic study of non-market decision making" (Mueller, 1979). Thus public decision making takes place within institutional and ideological frameworks, so that it impinges on and is affected by governments, politicians, bureaucrats and various interest groups. In this sense the study of public-choice economics has fully restored the title of 'political economy' to the discipline. More importantly, however, the public-choice approach explains, as Arye Hillman points out, why the maximization of welfare is not achieved. There are a number of factors that account for this. To mention just a few: insufficient information at the hands of the government for adopting correct policies; the correct policies may not be adopted because they are not in the interest of the government; those taking decisions, such as politicians and bureaucrats, have their own personal objectives; and institutional constraints and circumstances may prevent the adoption of certain desirable policies. The above suggests that while "the orthodox model provides a normative perspective where the focus is on what ought to be....the public choice model predicts that politicians' decisions may not result in the recommended policies being chosen and implemented. This is a positive proposition with no approval given or intended of political behavior" (Hillman, 1996; Olson, 1983). In this respect the study of the adoption of new economic policies in post-socialist societies in the 1990s is particularly instructive (Hillman, 1994).

Let us now go back to the 1960s and consider the outcomes of government intervention which was adopted in many Western countries. This worked well for a while, yet by the 1970s a new problem arose, that of 'stagflation'—the simultaneous existence of both high levels of inflation and high unemployment. Monetary policy seemed to work only for a while. The reason is that people cannot be fooled all the time: sooner or later they realize that higher money wages does not mean higher *real* wages in an inflationary situation; hence they press for higher real wages, which adds to the inflationary spiral and makes it not worth while for the entrepreneur to increase the number of workers he employs. Furthermore, much depends on people's expectations, which if exaggerated could further accelerate the inflationary trend. Keynesian economics seemed now to be in trouble. The crux of the problem for economists seemed to be how to explain that institutional 'corrections', such as advocated by the Keynesians, need not necessarily bring the economy back to stability. It seemed that some factor or factors were missing in the equation. Modern American economists during the past three decades or so have pointed increasingly toward one such factor—namely, the expectations of economic agents, be they consumers, producers, government officials, stock market speculators, or economists working as experts. The question was how to assess the expectations of such agents, for example regarding the future inflation rate. Thus, Milton Friedman and Edmund Phelps held that such expectations were based on past experience, hence in the short run monetary and fiscal policy could outwit economic agents, for instance by inducing a higher inflation rate than in the past. Producers would, as a result, wrongly assume that the price of labor was lower in real terms, and that they could charge more, in real terms, for their products. Similarly, workers would erroneously believe that their higher money wages actually increased their standard of living. But this could work only for a while. As we noted above, eventually all concerned realize that they had been fooled, and thus the economy returns to a 'natural' rate of employment/unemployment. This is because in the long run "changes in the money supply do not affect real variables such as employment and output after an adjustment period" (Miller and Pulsinelli, 1985).

A better way of assessing the expectations of economic agents is to take into account their future rational expectations, rather than extrapolating from past experience. This was proposed by J.F. Muth in 1961 and by Robert Lucas in 1981. Such a forward-looking assessment would be based on all available data, information about all true variables at work including government policy, such as the announcement of a strict monetary target and the determination to adhere to it in order to maintain credibility (*The Economist* 1995). The gathering of information and its analysis is not much of a problem. The sophisticated tools developed by econometricians handling the empirical side of economics can provide the necessary information (Marschak, 1994). What determines, however, the predictive power of such information is the 'correctness' of the economic model which serves as the basis for the econometric data and calculations. Since the theory of Lucas is based on *rational* expectations, this suggests that such a model would be the same as 'perfect foresight', except for random errors typical of all statistical data. What is characteristic of both the Milton Friedman[1] and the Robert Lucas[1] approaches is that they suggest a return to the basic assumptions of classical economics: a return to the notion of a self-regulating 'natural' employment/ unemployment equilibrium, as well as a return to the assumption of *rational* economic behavior.

The renewed support for the acceptance of the basic assumption of rational behavior is reflected also in the application of 'game theory' to economics. Game theory describes and explains situations in which participants adopt strategies based on optimal behavior, so as to achieve the greatest advantage for themselves. The participants will take into account the rules of the game, say those of a free competitive market, and each participant realizes that the outcome of his actions depends also on the interests and actions of the other participants. Thus, a producer is aware of the importance of how other producers, consumers, the government and economic agents such as banks, will behave in the given situation. "Each decision-maker is 'rational' in the sense that he is aware of his alternatives, forms expectations about any unknowns, has clear preferences, and chooses his action deliberately after some process of optimization" (Osborne and Rubinstein, 1994; Morgenstern, 1968).

Taking into account the strategies of optimization and the rational expectations is undeniably important for the study of economic processes and the outcomes of economic policies. However, economics as a science runs into the difficulty of how to cope with the psychological elements in the very *expectations* themselves. Thus, can economists, who rely on rational expectations, deny that, so far as entrepreneurial decisions are concerned, "the future is uncertain and expectations are volatile. Moods, the state of confidence, and the 'animal spirits' of the entrepreneurs play a key role in the formation of their expectations and, therefore, in investment decisions. The levels of activity and employment depend on imponderable, uncontrollable, and extremely unstable psychological factors" (Scrapanti and Zamagni, 1995).

Is the importance of the psychological factor in economics exaggerated? Does this factor destroy the basic assumption of rational behavior? Kenneth Arrow, another Nobel Prize winner, relying on experimental studies in cognitive psychology, which tested the abilities of human beings to make rational choices and form probability judgements rationally, points out that even in far less complex experimental situations than the real economy, the evidence suggests that the assumption of rational behavior is strongly contradicted, and that certain systematic kinds of bias are found: one is that people tend to overrate their own abilities, they show overconfidence in their judgments; the other is that choices which are objectively the same are looked at differently because of 'framing' which "is controlled by the manner in which the choice problem is presented as well as by norms, habits and expectancies of the decision maker" (Tversky and Kahneman, 1988).

Arrow also adduces evidence from the securities markets where "stock and bond prices fluctuate far more than is consistent with rational behavior on the part of the participants". Let me quote in full the conclusion of this prominent economist: "The image of the future, cloudy though it be, powerfully influences the current state of the economic world. The formation of this image owes something indeed to our individual and collective efforts to use our experience and our preconceptions to that end. We are not without sense and reason, but we are subject to the necessity of oversimplification and to biases built into us. We can effectively use only part of the knowledge that is or could be available to us. But even if we used all we could, our prevision would be deficient because so much can happen that will necessarily be a total surprise" (Sichel, 1989). Economics, treated in the foregoing, concerns the science developed for the understanding of the workings of free or partly-planned economies; not of tightly controlled and totally planned economies.

Considering the uncertainties and complexities of the economic world we presented, and also the questions concerning the rationality of human actors in the economy, it is no wonder that Max Planck, the famous physicist, is reputed to have remarked to John Maynard Keynes that "he had once thought of studying economics but had abandoned the idea because he found the subject too difficult". Planck's remark is usually interpreted as meaning that the more precise subject of physics had a greater appeal for his mind than the uncertain human-dependent subject of economics (Seger, 1972).

The Cognitive Factor—a Crucial Determinant?

The account of developments in economics suggests that the subject has returned to its basic micro-economic assumptions of rational choice and rational expectations, exercised by economic agents wishing to maximize their satisfaction (utility). A major criticism is that this has not really helped in the prediction and explanation of such matters as business cycles, economic development and stagflation. Does this vitiate the standing of economics as a science? We can follow this with the rhetorical question: "If economics isn't science, what is it?" (Rosenberg, 1994). The appropriate comment is that much depends on how science is defined. If science requires exact reproducible observations and fully controlled experiments to provide explanations and predictions of natural events, then obviously economics cannot meet such requirements. However, if the label 'science' can be attached to a discipline where regularities are shown to exist and where a long-run prediction of the probability that certain steps will have certain consequences can be made, then economics may be regarded as 'scientific'. The same applies, of course, to all the social science disciplines: sociology, political studies, social psychology, criminology, and so on.

This wider and looser definition of the 'scientific', when applied to human behavior, appeared to be necessary because of the *double-cognitive bind* of the social sciences. Not only does the social scientist—like the natural scientist—rely on his cognitive powers of perceiving, thinking, imagining, reasoning, conceptualizing, judging and drawing conclusions; the social scientist's domain of study consists basically of actions by conscious human beings with their own cognitive make-up, which is not the case in the natural sciences. Many social scientists have come to appreciate this fact to such an extent that they have turned increasingly to psychology in order to solve the problems in their own disciplines. Some economists, as for instance Kenneth Arrow whom we quoted above, have acknowledged that the basic simple assumption of rationality is problematic, in view of some experimental work in psychology showing that "individual behavior systematically violates rationality" (Frey, 1992). Since human beings are capable of acting both rationally and irrationally, it has been argued that there is a need to find out how the acting individual perceives the situation in which he acts and his state of mind in the particular situation, something that only psychology can do. Consequently, "psychology has been drawn upon to explain fads in buying, and stock market cycles", and it has also been applied to "other economic phenomena that seem to be poorly accounted for by the standard behavioral assumption employed by economics" (Mueller, 1994). Economists like Herbert Simon (another economics Nobel laureate) have adopted a cautious stance by using the concept of *bounded rationality*. This "stresses that an individual has only limited computational abilities when it comes to making a decision. Simon assumed that economic agents can never make truly optimal decisions because they are inherently limited in the amount of information that they can process prior to having to make a decision" (Casti, 1992). The tendency in economics to turn to psychology has been criticized from the anti-reductionist point of view. Although underlying economic processes there are psychological elements such as preferences and strategies, the economic and the psychological are not coextensive domains. As Jerry Fodor argues, therefore, "to reduce economics to psychology would....involve a very great deal more than showing that the economic behavior of groups is determined by the psychology of the individuals that constitute them" (Fodor, 1976).

In sociology there has been a good deal of resistance to the reduction of the discipline to psychological explanations. But even among sociologists, many have argued that it is essential to understand the subjective states of individuals and the meanings invested in their acts. They obviously do not deny the importance of social interactions and the existence of social structures, but insist that the latter are outcomes of individual action based on behavioral principles (Mennell, 1974). Such principles, in the view of George Homans, rest on the idea that rewarded action will be repeated, the obverse of which is the avoidance of or escape from punishment (Homans, 1982). Raymond Boudon represents a less psychological standpoint, but he, too, stresses that "individual action and the products of individual action constitute the only and ultimate reality with which the sociologist has to deal with". Boudon, however, emphasizes that these individual actions take place in situations of interaction, which are variable and immensely changeable. It is for this reason that general laws, trends and regularities have "so many exceptions, that one wonders in which sense they are still rules". Hence, Boudon goes for *ad hoc* explanation in the singular micro-context of the "situation of interaction" (Boudon, 1983).

The foregoing examples from economics and sociology show the awareness of many social scientists of the crucial role played by the thinking and acting individual in the dynamics of the economy and society as a whole. However, the importance attached to the psychological element stems in many cases from the desire of social scientists to make their disciplines more truly scientific. They see their salvation in explanations based on the analysis of individual cognitive behavior. There are two reasons why I do not think that this will advance the social sciences. While there is no denying that cognition underlies all meaningful acts, this cannot do away with the fact that the acting individual is influenced by the wider context and the long-lasting macro-phenomena, such as the economic, social and political structures developed by a particular society, its culture including language, its history, its norms, scientific ideas and developments, and its relations with other societies. Yet, knowledge of the nature of the cognitive process in the individual will not enhance our understanding of the dynamics of the societal structure built up by ongoing human social life. Let me exemplify this: An individual is in conflict with another individual, a situation in which they are exchanging ethnic prejudices. Now, there is no denying that in their strategies they will take into account how widespread ethnic prejudices are in their pluralist society and what resources, power and status are at the command of each ethnic group. Of course, in analyzing such a conflict and its outcome we can take into account all the 'givens' of the situation, including the measure of ignorance of the individuals involved regarding the social or political conditions in their society, their ideologies, their misconceptions, their strategies, as well as their individual cognitive processes taking place in such a situation. However, I cannot see how more knowledge about the cognitive processes in the minds of these individuals can provide us any knowledge concerning the macro patterns existing in the society where the conflict situation occurs. Yet it is the latter that we have to apprehend and understand, something we cannot do without the observation and aggregation of the behaviors of many individuals, and without an account and analysis of the institutional and structural aspects of society.

The second reason for my belief that an appeal to cognitive psychology, and the reduction of economic and social phenomena to psychological building blocks, will not help matters, is that cognitive psychology tends to push us towards neuroscience and thus away from what is characteristically *social* about human behavior and human interaction. The reference to neuroscience relevant here is the neuro-psychological investigation of complex mental activity by the analysis of the internal structure of mental processes and neuro-psychological research concerning the foundations of personality and behavior (Luria, 1973, Vernon, 1994). Now, let us for the moment agree that individual human intentions, desires and beliefs *are* important as building blocks for understanding the relations between human beings and, therefore, for the outcomes of such relationships, seen in the institutions and structures of society. If such intentions, desires and beliefs place people into *types* of mental state, can we identify such states by the observation of behavior or by neuroscientific identification of mental states? This is very doubtful. The basic mental state that neuroscience tries to identify and understand is consciousness, yet there is no agreement even as to how to define it. Thus, it is not clear as to how to incorporate consciousness into science.

There is no doubt that conscious experiences exist, whether it is pain one experiences, seeing the sunset, having feelings of moral indignation, or being intent on taking some action. Conscious experience is subjective reality and it gives the clearest expression to what is called "the mind"; yet even interdisciplinary efforts of neurobiologists, cognitive psychologists and philosophers have not produced an agreement as how best to study consciousness. While it is agreed that consciousness is somehow based on the workings of the central nervous system—that is, the brain and the spinal cord in their coordinative function—the mind-brain relationship is far from being clearly understood. Neurophysiological explanations based on identifying consciousness with some as-yet unknown neural system has not been met with universal acceptance "because it seemed that you cannot identify something mental, like consciousness, with something nonmental, without already leaving out the mental". It follows that consciousness and intentional behavior may "not be strictly identical to any specific type of physical matter" or physical process (Revonsuo and Kamppinen, 1994). As Alexander Rosenberg maintains, in the case of intentional behavior we cannot break into "the intentional circle" by studying, for instance, bodily movements. And he adds that "the brain states neuroscience identifies just don't seem to line up in lockstep with mental states that introspection reports" (Rosenberg, 1994). In other words, there appears to be a discrepancy between the actual mental phenomena and neural reality (Revonsuo and Kamppinen, 1994). Furthermore, the complete identification of the mind with the brain creates a new brain/body dualism. That is to say, if thought processes, consciousness, emotion are seen as centered in the brain, the latter being the 'seat of consciousness', then the continuity between mind-brain and body tends to be denied, ignoring the role of the body through sensory stimuli and the sheer physiological functioning of the body in affecting consciousness. So, even a materialist approach has real difficulty in precisely identifying and locating that infinitely complex aspect of human existence which we call consciousness (Pepperell, 1995; Polkinghorne, 1998).

The extreme difficulty of tackling the phenomenon of consciousness scientifically has led some scientists to deny its very existence. Michael Polanyi quotes and ridicules such extreme empiricists who deny the existence of consciousness, because it is not a tangible external fact (Polanyi, 1974). Others have attempted to explain consciousness by means of simulated computational activity. Roger Penrose, a leading scientist in the field of consciousness, admits that, at this stage, it is impossible to provide a precise definition of consciousness; but he states clearly that it is *something* that transcends computation. Hence we should correct the analogy between the computer and the mind, an analogy in which things have become inverted. The mind is not like a computer, since it has feelings and intentions, but rather the computer is in some respects (computation, memory) like the mind. Penrose's stand is clear regarding the particular mental quality of *conscious understanding*, namely "that it is essentially impossible that such a quality can have arisen as a feature of mere computational activity, nor can computation even properly simulate it" (Penrose, 1994). His conclusion is "that whatever brain activity is responsible for consciousness....it must depend upon a physics that lies beyond computational simulation". Penrose is a reductionist who believes that the mental state of consciousness will be eventually explainable by the physical processes in the brain (Penrose, 1994).

On the whole it seems that we are a very long way from clear scientific statements as to what consciousness is, and how it arises. We do conceptualize it intuitively in a wide sense as expressing human awareness, human will and human feelings. The shortcomings[2] in this area of science suggest that our understanding of intentional behavior will remain, at least for some time to come, at the interpretive level, and that such information, when useful, must be integrated into our knowledge regarding the workings of the wider framework of the humanly developed *social world*, which is the real subject matter of the social sciences. So, while in the abstract, and fundamentally, the cognitive factor *is* crucial, since no social world would exist without the conscious acts of thinking individuals, in realistic terms it is the social network and its dynamics, quite apart from its creative intentional individuals, that the social sciences must observe, analyze, and understand.

NOTES

1 Both men are Nobel laureates.

2 Note that the subtitle of Roger Penrose's book is: *A Search for the Missing Science of Consciousness*.

CHAPTER SEVEN

THE SCIENTIFIC WORLDVIEW

In the previous chapters I looked at a number of aspects of scientific work, the endless activity that produces knowledge. This knowledge making is always fragmented, and it always leads to further questions which test our ability to get to know the world around us[1] better—but probably never fully—by accumulating more knowledge. Scientists certainly cannot predict what knowledge will be added in the future. Will particle physics ever answer the basic question of what is the origin of mass or how is it generated? Will economics come up with a satisfactory formula for preventing business cycles?[2] Scientific work does not add up to some kind of a 'whole', an integrated grand science. Yet, there is something that holds all scientific work together: this is a particular worldview which provides the basic assumptions according to which scientists work. The English word 'worldview', freely translated from the German *Weltanschauung* and *Weltansicht* means a number of things: a comprehensive conception or image of the universe, man's relation to it, and how reality is interpreted. The scientific worldview can be labeled as the 'super-paradigm' of modern science, the framework and rules which all scientists are supposed to accept and follow. This worldview also makes a clear distinction between 'scientific' and 'non-scientific' knowledge.

Scientific Knowledge Supreme

The worldview held by most scientists is that scientific knowledge is superior to all other kinds of knowledge. This is so, it is argued, since science is based on rationality. Furthermore, a basic requirement in science is to obtain purely factual information and to analyze such data by the best technological means and the accepted methods, such as experiments and statistical checking, which ensure that the *true* state of affairs is presented concerning the reality in which we live. Therefore, when science is compared to other kinds of human knowledge, such as produced by religion, folklore, art and literature, the supremacy of science should become obvious.

The view that science is supreme in the realm of knowledge and, above all, that it provides correct knowledge has prevailed throughout the twentieth century. Bertrand Russell (1872–1970), the renowned British philosopher and mathematician, who was also a winner of the Nobel Prize for literature, has contributed, through his prolific writing and eminence, probably more than anyone else to the upholding of such a supremacist scientific worldview. Russell drew a sharp distinction between the pre-scientific outlook and the scientific one, and hailed the latter's triumph over religion, which relied on unproved belief, on authority and tradition. Science, on the other hand, relied on empirical knowledge; experience is thus the foundation of all our knowledge of the physical world. By applying logic to such experience, we acquire a knowledge of natural laws, resulting in a corpus of knowledge which provides the correct account and explanations of the physical world. Russell held the view that "the power of science has no limits", although he tempered this by saying that "the pronouncements of science do not claim to be certain, but only to be the most probable on present evidence". He saw the progress of science in that each age is more scientific than the previous age: "if our ancestors would consider present society very scientific, it is likely that to our descendants it will seem the reverse" (Russell, 1976; Russell, 1931).

A much more attenuated worldview of science is that presented by Michael Polanyi. For him the facts on which science relies do not accrue in a detached sort of way, so as to provide a solid, humanly untainted basis for scientific knowledge. The facts of science are obtained through *human* experiences. This means that we must discard a rigid conception of science by incorporating into our conception of scientific knowledge "*the part which we ourselves necessarily contribute* in shaping such knowledge". Therefore Polanyi embarks on a critique of the so-called exact sciences "in order to displace quite generally the current ideal of detached observation by a conception of *personal* knowledge". Thus, there can be no rule, he says, on which we can rely "for deciding whether the discrepancies between theory and observation should be shrugged aside as observational errors or be recognized, on the contrary, as actual deviations from the theory. *The assessment in each case is a personal judgment*" (emphasis added). He points to evidence that even the most highly automated instruments for recording physical facts cannot avoid the uncertainty attributable to the personal bias that could affect a series of readings by the scientist. "Even the most exact sciences must therefore rely on our personal confidence that we possess some degree of personal skill and personal judgment for establishing a valid correspondence with—or a real deviation from—the facts of experience". So, uncertainty creeps in because ultimately we must rely on our judgment as to whether we are perceiving the facts correctly, and whether the recorded facts are in agreement with accepted theory. Even the slightest gap between theory and instrument readings produces that "thin end of the wedge" which completely separates personal knowledge from detached objectivity. "Personal, tacit assessments and evaluations...are required at every step in the acquisition of knowledge—even 'scientific' knowledge" (Polanyi and Prosch, 1975). Yet Polanyi does not claim that because "the *personal participation* of the knower in all acts of understanding", knowing and understanding become totally *subjective*. On the contrary, such knowing is *objective* "in the sense of establishing contact with a hidden reality". It is this fusion between the personal and the objective that produces scientific knowledge, which he calls "personal knowledge" (Polanyi, 1958).

The view that science is supreme, in the sense that it has a special status as compared to various other kinds of knowledge, has undergone more sustained criticism by some modern sociologists, interested in the sociological study of science and knowledge as a whole. Earlier sociologists, however, like Karl Mannheim, Werner Stark and Robert Merton accepted the standard view of scientific knowledge, which denies that scientific knowledge is *socially* contingent. That is to say, the substantive laws of nature that science discovers are said to have their independent veracity quite apart from the individuals, social institutions or societies by whom and in the context of which they happened to be uncovered and confirmed. Furthermore, these sociologists were careful to apply this standard view to the natural sciences only. They held that the arts and the social sciences deal with realities whose meanings are created by human beings, whilst natural science identifies realities which already exist in nature (see Note 2, this chapter).

At this point it is worth quoting at some length from the succinct presentation by Michael Mulkay of what the standard view of science is, in order to appreciate fully the claims that scientific knowledge is something special and supreme in the realm of human knowledge. Subsequently some criticism will be leveled at these claims.

> From the perspective of the standard view, the natural world is to be regarded as real and objective. Its characteristics cannot be determined by the preferences or intentions of its observers. These characteristics can, however, be more or less faithfully represented. Science is that intellectual enterprise concerned with providing an accurate account of the objects, processes and relationships occurring in the world of natural phenomena. To the extent that scientific knowledge is valid, it reveals and encapsulates in its systematic statements the true character of this world....Although the natural world is, in a certain sense, undergoing continuous change and movement, there exist underlying and unchanging uniformities. These basic empirical regularities can be expressed as universal and permanent laws of nature, which tell us what is always and everywhere the case. Unbiased, detached observation furnishes the evidence on which these laws are built... Indeed, observational laws are no more than general propositions summarizing a body of reliable factual evidence. The validity of the factual foundation of scientific knowledge can be guaranteed with a high degree of confidence because science has evolved stringent criteria, for example, in connection with experimental procedures, by means of which empirical knowledge-claims are evaluated and their accurate representation of empirical phenomena is ensured. Thus accepted scientific knowledge, because it has satisfied these impersonal, technical criteria of adequacy, is independent of those subjective factors, such as personal prejudice, emotional involvement and self-interest, which might otherwise distort scientists' perception of the external world. (Mulkay, 1979)

One of the salient points in the 'standard view' is the uniformity of nature, meaning that there is some kind of regularity and constancy in the natural world and consequently the basic laws uncovered apply always and everywhere. But can this be proved empirically? Mulkay refers to Hanson's pungent argument that we cannot avoid a vicious circularity. If we wish to test the principle empirically without assuming that uniformity exists, then particular observations will not decide the issue. If, however, we make the assumption of uniformity so that particular observations *are* sufficient for us as evidence, then we have not actually tested the principle itself. As Mulkay puts it, "Unless the principle is assumed to operate, there is no way in which we can infer it from particular observations" (Mulkay, 1979; Hanson, 1969). Now, since we do not have sufficient evidence from the entire universe, it is conceivable (if we do not assume uniformity) that the 'law of gravitation' does not apply everywhere in the universe. The 'law of gravitation' therefore expresses the acceptance by scientists of a theory of gravitation based on existing evidence—but it is no proof of the uniformity of nature.

Another important claim of the 'standard view' is the validity and accuracy of the facts adduced concerning the natural world. Scientific facts seem to be extensions of common sense facts, in so far as the same logic and consensus can be applied to the matter-of-fact world both in everyday experience and in scientific work. Whether laymen or scientists, we are all aware of the difference between solids and liquids, between things in motion as against stationary things, and so on. As John Ziman says, *common sense realism* is universal, but this "is not due to all human minds having been genetically created as identical perceiving communicating instruments like remotely controlled television cameras landed on Mars. We derive our well-founded confidence in the reality of everyday things from the sensorimotor bodily experience of childhood, reinforced and brought to consciousness by linguistic interaction with other people." So, our apprehension of everyday facts as well as of facts necessary for acquiring scientific knowledge is dependent on individual consciousness and socio-cultural contexts (Ziman, 1978; Ziman, 1984).

A problem arises as soon as we dig a little deeper into the way the facts are established. The method is that of representing the object which is being apprehended. This means that we create an image of the reality we perceive, and the question is how can we be sure that the former truly reflects the latter: is our representation of reality valid? How do we know, for example, that in the work of a solid-state physicist there is a true correspondence between the inscription of a pen chart recorder and the current state of atomic alignment in a sample of metal alloy? That is, how do we know that what is being recorded is a true reflection of the system or process investigated? As Steve Woolgar points out, this problem "pervades every aspect of research practice or inquiry". There are three aspects to this problem: First, representation is carried out by some form of communication; but however precise the latter, be it even mathematical symbols and formulae as preferred in the natural sciences, it is the *meaning* it conveys, that is the content of the communication that matters. And since different meanings can be ascribed to the same event, it is the particular vantage point or purpose of the apprehension of facts which dictate the meaning of the fact observed. In this sense there is *no one true* representation. Second, attempts to define precisely the representation of the facts may be endless, since "it is always possible to ask for further clarification, elaboration, elucidation and the like". And third, there is an intricate two-way relationship between the object observed and its representation. The fact observed triggers and sustains its representation, yet the representation characterizes the nature of the fact observed. The interpretive element in adducing scientific facts cannot be eliminated (Woolgar, 1988).

Even at the technical level, achieving accuracy in representing the facts of nature is a relative matter. Technological refinements, experimental precision, and the steps taken to eliminate any subjective elements will achieve a degree of accuracy only from the perspective of the particular purpose for which the fact is being represented. Thus, does a photograph of a person, an object, or a landscape, capture more accurately and on the whole more truly the reality it represents than a painting produced by an artist? After all, the photograph is based on scientific principles which we assume have no subjective elements in them, while the painting is obtained through the personal and subjective perceptions and judgments of the artist. The distinction, however, is not that clear-cut. Although the photographic process is scientifically based, the camera and other instruments used in carrying out the process and producing the photograph are manipulated by a person; and even if he is a scientist he will use his particular powers of perception and judgment in carrying out the act of photographing. Furthermore, it is conceivable and quite probable that the artist, using his inborn talent rather than scientific rules, can capture certain subtle characteristics of what he paints which the most sophisticated camera cannot do. It is, therefore, not a question of a 'better' or 'truer' representation, but rather the purpose and vantage point which is at stake when capturing and representing the facts in our world.

Finally, as we have seen, the 'standard view' incorporates a clear empiricist stand, which means that scientific knowledge is held to be "the sum of all those propositions which have survived generations of repeated confrontation with reality. Empirical testability is the mark of the openness of claims about the world to revision under the impact of the world itself" (Benton, 1977). But this empiricist conception of scientific knowledge is fraught with serious difficulties. Strict empiricism holds that we must clearly distinguish between theory and observation. It is the observation of facts which leads to the confirmation of laws proposed by the theory. In other words, the theory becomes scientifically legitimate by reducing it to observable consequences. But this is not at all so simple. In order to apply the theory to the observable world we must use some kind of bridging concepts which are partly theoretical and partly empirical, as for instance the concept of kinetic energy used in the kinetic theory of the behavior of gases. But kinetic energy is a composite term: that is, the motion can be observed, seen in the pressure exerted by what we consider to be the constant collision of molecules depending on the temperature of the gas, but energy as such, even if we can calculate its amount, still remains an unobserved element.

Of course, there will always be unobservable aspects of reality—the hidden parts of nature (see Chapter 3, pp. 41–42). And it is also true that what *is* observable is theory-laden. We cannot attribute meaning or significance to our observation without putting it into some conceptual framework: that is to say, theory-neutral observation cannot be achieved either in day-to-day life or in scientific investigation. Thus, scientists' observations in the laboratory are based on a selection of scientific knowledge acquired in the past, and the results which accrue from the laboratory experiment set the scene for the forthcoming scientific observations, which will undergo interpretation in the light of already existing scientific theories (Woolgar, 1988; Ryle, 1949; Mulkay, 1979).

Let me carry a little further this discussion of whether science is firmly based on observed facts by making two more points. First, the traditional distinction between observable and unobservable entities is problematic. For example, are quasi stellar objects, known in science as 'quasars' observable? The answer is both yes and no: we do pick up X-ray emissions from very distant radio sources thought to be violently active cores of distant galaxies, but we cannot observe quasars as such (Uvarov and Isaacs, 1993). The second point is, however, that some of the hidden aspects of nature may prove to be only temporarily hidden. For instance, predictive theories hypothesizing some hidden mechanisms have often led scientists to the discovery of observable patterns, as in the case of the idea of a 'genetic code' which came to be shown to be carried by strings of chain-like molecules of DNA (see Chapter 2; also Ziman, 1978).

Much of the foregoing has highlighted the skepticism concerning the claim that science is based on pure facts revealing the uniformity of nature and its fundamental laws. Empiricist science can, however, be buttressed by its applied aspects. If its claim that its foundations rest on objective facts appears dogmatic and is open to criticism, in its application to the real world science is demonstrably factual, so far as its results or consequences are concerned. Producing electricity in nuclear power plants and bringing about new human life by in-vitro fertilization are not imaginary events. But this 'proof of the pudding is in its eating' argument, attractive as it is, may prove to be double-edged. If the application of scientific theory can alter the course of events, does this not show the opposite of a uniform reality? Furthermore, since scientific theories may be falsified by new evidence, the laws proposed by such theories cannot be said to be immutable. In his oft-quoted metaphorical description of scientific knowledge, Karl Popper tells us that there is nothing 'absolute' about objective science. Science "does not rest upon rock-bottom....The bold structure of its theories rises, as it were, above a swamp. It is like a building erected on piles. The piles are driven down from above the swamp, but not down to any natural or 'given' base; and if we stop driving the piles deeper, it is not because we have reached firm ground. We simply stop when we are satisfied that the piles are firm enough to carry the structure, at least for the time being" (Popper, 1959). Scientific knowledge contains elements of uncertainty since it is not based on unchallengeable facts. As we shall argue below, these facts of reality are too complex and the reality of the universe is ultimately incomprehensible to the limited cognitive abilities of the human race producing scientific knowledge. Again, such knowledge and the scientists themselves are never detached from their social context. The advantages, superiority and factual firmness of scientific knowledge are, therefore, expressions of a worldview which competes and challenges other worldviews, all of which may be described as 'belief-systems'.

Science as a System of Beliefs

Science has a number of desiderata for producing scientific knowledge. These desiderata involve *a priori* propositions—that is, truths determined and defined as such without reference to experience. An example may be assumptions concerning the existence of certain patterns and laws in nature. Another *a priori* assumption is that patterns of logical relationships exist in reality, and that therefore logical reasoning and logical constructions are the ways in which we arrive at the *scientific* understanding and explanation of real events in the world. A technical desideratum, but one with strong implications for the view that scientific facts arising out of research truly represent reality, is for scientists to take decisions, based on their experience and judgment, as to the best way of communicating unambiguously their empirical findings. Another facet of science is its use of categories of thought and language and its need to classify things, to isolate systems, and typify phenomena. Finally, it is assumed in science that the human mind can fathom the workings and secrets of the existing world.

Taken together these requirements form *a system of beliefs* which is strongly upheld in science. A brief examination will show that these basic requirements tilt the purely empiricist foundation of science. As a result either an inconsistency arises as far as naive empiricism is concerned, or science and its brand of knowledge must be defined in wider terms as the product of human confrontation and imagination with the real world.

Let me elaborate on these points. The first issue mentioned concerns *a priori* propositions. Even Bertrand Russell, who consistently held that experience is the foundation of all our knowledge of the physical world, admitted in later writings that *a priori* knowledge was needed, in addition to experience, in order to gain scientific knowledge (Sainsbury, 1979; Narkhnikian, 1974). In a similar vein Max Planck (1858–1947), the famous German physicist and Nobel Prize winner for physics, who saw the exact sciences as based on facts acquired by sense experience, nevertheless declared that scientific thought and research cannot be without presuppositions at all (Planck, 1949). I shall exemplify the ensuing dilemma for empiricist science by looking at its basic assumptions concerning the essential and inherent characteristics of the material world. It is true that such assumptions have changed in modern science in view of new findings and the interpretations of such findings. Thus, deterministic views have given way to probabilistic ones, and, according to some, nature which was held to be orderly actually shows chaos. But it is not the exact postulation that is crucial here. Rather, it is the very idea or the basic assumption that there is *an underlying pattern which is universal*, that is undeniably a matter of belief.

Secondly, we have to raise the question whether it is purely logical reasoning that enables scientists to see causal connections in the world—this would support the claim of objectivity—or whether there are extra-logical elements in human perception, memory, cognition and intuition which play important roles in arriving at scientific explanations—thus requiring that subjectivity be retained in science. If both these facets are there then they are bound to produce a gap between the claimed logical foundation of science and real science. The idea of logically acceptable structures, supported by the prevalent belief that there are "clear structures behind complex events" (Feyerabend, 1991), appears to be something that science and philosophy impose on nature in an attempt to bridge the gap between actual experiences and their interpretations. These logically acceptable structures then dictate the sorting out of scientifically pertinent facts and the acceptable interpretations, and thus produce legitimated or accepted scientific knowledge.

Next, there is the issue of transmitting information that science acquires about reality and the need to analyze it in the clearest possible way. For this purpose various means or methods are used: mathematical formulae, physical and pictorial models, and verbal presentations. Two points may be noted here: First, that even the use of the most unambiguous language such as mathematics, which is based on axiomatic or self-evident truths and which is a medium very well suited for a subject like physics, does not ensure that the message communicated is meaningful or true (Ziman, 1978). It is not the absolute logical consistency of the means of communication that matters, but whether such a system of thought represents the true empirical relationships in nature. Second, the selection and use of what appears to be the most appropriate means of communication or method of analysis for the particular area of scientific enquiry is ultimately based on the decision of the scientist. His judgment stems from a mixture of his experience in confronting the empirical realities of his subject-area and his belief in what are the best ways of dealing 'scientifically' with the confronted reality.

Turning now to the abstract categories of thought—the concepts about reality such as time, space, the mind, and the classifications of the concrete contents of the world perceived—we shall find that, like common sense conceptualizations and classifications, so in science, too, these are based on cultural developments embedded in societies and languages. They are not neutral. The question is whether science can neutralize the cultural and historical effects on its systems of classification and its repertoire of concepts. If it cannot do so it is possible that science produced in one culture will differ from science produced in another culture. John Ziman has this to say about medicine developed in different cultures: "In clinical medicine....there are such large differences of categorization and interpretation that Western and Chinese doctors might appear to be talking of quite different physiological phenomena. Having conceded that Western medicine is far from complete and consensual, and that it may have as much to learn from Chinese theory and practice as it can pretend to teach, we must note this disturbing evidence that the scientific investigation does not converge inevitably on the same, unique, 'objective' map" (Ziman, 1978). Science develops, through historical stages and in different societies, often on different tracks and in different directions; it is not culturally neutral and it is not universal, although the spread and dominance of modern Western science, through its tremendous success, tends to push it towards universalization.

Finally, there is the question of the capacity of the human *mind*, if a dualist position is taken, or the human *brain*, if a monist one is taken. In any event, there can be little doubt that it is through the intuition, insight, consciousness and overall capacity of the human brain/mind that scientific knowledge comes about. There must be a knower, and this is where science begins and ends. Even the most sophisticated computer can produce knowledge only through human involvement. The most advanced computers that can accumulate and process huge amounts of information and analyze a tremendous variety of combinations of information, to degrees unimaginable only a few years ago, would still help to produce knowledge limited by what the human mental capacity can understand. The underlying assumption of science, so far as the human brain/mind is concerned, is that it has the capacity or "the powers to discern coherence in nature" (Polanyi and Prosch, 1975). Granting that human beings have such mental powers, their limitations must be accepted. So far science has not come up with evidence to show that the human brain/mind capacity is able "to understand experience completely and in every respect" (Lewis, 1972). On the contrary, both scientists and philosophers hold that our mental capacities have strict limitations. To quote John Ziman, a physicist: "The key factor in the evolution of science remains....the power of the human mind to recognize a comprehensible pattern in a mass of detail. This faculty has its limitations; we may reasonably ask whether scientific knowledge itself is fundamentally restricted by these limitations" (Ziman, 1978). The problem looms even larger if we consider not only the limitations of *homo sapiens*, but also the infinite complexity of the world they face as scientists. On this point Christopher Cherniak, a philosopher, notes that "the finitary predicament", that is the "fixed finite limits of our cognitive resources" is the greater and most subduing if we take into account the universe's inhumanly complex nature. To understand fully this complexity the scientist will face an intractable situation: considering the extremely short span of life and the limits of mental capacity even just the information processing becomes unmanageable because "the number of possible combinations among the elements explodes exponentially". In sum: "The breadth and depth of putatively possible knowledge may be intrinsically too great for a both manageable and complete world view..." (Cherniak, 1986). I am using and adapting Cherniak's points to press the argument that science has no privileged position for achieving complete and apodictic knowledge, that which tells us all there is to know about the world and which is clearly demonstrable and established on uncontrovertible evidence. Science, after all, is built on human existence and human beliefs, which simply means that it has clear limitations.

My view of science as constituting a system of beliefs, and my reference to it as an entity, have served as abbreviations for what sociologists would define as the interactions of and the resulting knowledge produced by a collectivity or a community of scientists. Their tasks and aims to produce scientific knowledge have been institutionalized, the clearest expression of which is the ethos of science. Robert Merton, a prominent contemporary sociologist, defines this ethos as a "complex of values and norms which is held to be binding on the man of science".

Although the norms and values have not been codified they are expressed in practice as well as in writings about what are the right ways in science. Often there is also moral indignation directed against those contravening what is acceptable in science. As Merton puts it: "The institutional goal of science is the extension of certified knowledge", that is knowledge accepted by the scientific community. Here science has a number of prerequisites. Basically, it demands "empirically confirmed and logically consistent statements of regularities" and "sustained true prediction". These demands are binding "not only because they are procedurally efficient, but because they are believed right and good. They are moral as well as technical prescriptions". Merton also discerns four sets of institutional imperatives: One is universalism, namely that truth-claims in science, whatever their source, are subjected to preestablished impersonal criteria of objective confirmation. Another element is communal possession of scientific knowledge: there may be controversies about priority in some discovery or theory, which is even encouraged by the accent on and the esteem given to originality in the world of science, but "this does not challenge the status of scientific knowledge as common property". Competitive cooperation is encouraged, secrecy is frowned upon, open communication and the diffusion of knowledge is the norm, all in the interest of advancing the boundaries of knowledge; the results of scientific work being regarded as part of the public domain and important to the scientific fraternity. A third institutional element is that of disinterestedness. This is required for the truthful verifiability of scientific research results by the exacting but disinterested scrutiny of fellow experts. The accountability of scientists to the evaluation of their work by their peers, and their trusting in the disinterestedness of those checking on their work is conducive to a socialized sentiment and to institutional stability. Another norm is what Merton calls "organized skepticism". This is regarded as a virtue and is encouraged by the strict scrutiny of all scientific claims. When doubts appear concerning a theory or some findings it becomes legitimate to launch attacks against such theory or findings. This is in tune with the norm of the openness of the scientific enterprise to new challenges. In addition to the four basic institutional norms, Merton also discusses the norms and problems concerning plagiarism and the emphasis on originality in science (Merton, 1973; Ziman, 1984).

Some critics of institutionalized science point to the detrimental effects of such norms for science itself. As Paul Feyerabend argues, the attempts to enforce universal rules makes science into a dogmatic religious institution. He admits that dogmatism in science may, nevertheless, be necessary for its functioning, and so, I presume, are particular norms and myths. But Feyerabend argues that these rules are often unrealistic and that many scientists do not conform to them. For instance on the issue of skepticism he holds that the great majority keep it at a minimum: "it is directed against the views of the opposition and against minor ramifications of one's own basic ideas, never against the basic ideas themselves. Attacking the basic ideas evokes taboo reactions which are no weaker than are the taboo reactions in the so-called primitive societies" (Feyerabend, 1975; Redner, 1987).

A parallel has been drawn between science and religion: both constitute a kind of 'belief system', and both have an ethos and an institutionalized framework. Emile Durkheim believed that "these two forms of human mentality, however different they may be, far from deriving from different sources, are created one by the other, and are two moments in the same evolution". Durkheim based this conclusion on his account of the genesis of the most essential ideas of the human mind, which philosophers call categories of thought, such as time, space, force, causality, and "which dominate all logical activity". These were developed, Durkheim tells us, within the very center of religion, and "science has borrowed them from religion". Durkheim regarded all these essential ideas or 'categories' as social creations which could vary form society to society and from period to period. But although science and religion emerged from the same social sources, Durkheim saw science as a more advanced cultural product of society which in the modern world becomes a substitute for religion (Giddens, 1972; Mulkay, 1979).

Like Durkheim, who regarded science as part of a social evolution which enhanced objectivity, so Karl Marx, too, saw science from an evolutionary perspective. He regarded science and capitalism to have been at first liberating forces from the pre-capitalist religiously based societies. Yet with the development of capitalist society science became an exploitative resource: it was mainly financed by capitalist entrepreneurs and it produced knowledge and technologies which advanced their interests. The link between science and society was clear in Marx' view, just as it was in Durkheim's view. Some writers have also pointed to Marx' idea that "science generates its own ideology". Such an ideology would be 'scientism', a doctrine "which asserts that science, and *only* science can tell us what the world is really like"; an ideology which "has become so dominant in present-day industrial society that any knowledge-claim which falls outside its scope is widely regarded as vacuous" (Mulkay, 1979; Ziman, 1984). As Jürgen Habermas puts it: "'Scientism' means science's belief in itself: that is the conviction that we can no longer understand science as one form of possible knowledge, but rather must identify knowledge with science" (Habermas, 1978). Steve Woolgar refers to the "modern obsession with technical rationality and reason" as "the public face of the ideology of representation", which has been fostered by science. By this scheme of things the scientist is merely the agent who negotiates between the objects of the world and their representation; he is a passive and objective agent who simply reveals the facts of the world as they are (Woolgar, 1988). This idealized notion of science and many other pointers in this chapter suggest that there is a kind of ideology that one can identify with science.

The ideological dimensions of science and its dogmatism, relying on certain presuppositions and a particular worldview, has brought science into clash with other worldviews, science becoming in particular the chief antagonist of religion in the industrialized and progressively secularized modern Western societies. Robert Bellah has lamented this clash, arguing that the two worldviews can be integrated so as to furnish a fuller multi-faceted interpretation of reality as a whole, and thus enriching our understanding of total reality (Bellah, 1970). Bellah emphasizes that this is possible since religion and science represent different realms, the former interested in the meaning and significance of man's and the world's existence, the latter in the pragmatic technical explication of these existences. But both are fundamentally based on convictions, as are political ideologies, agnosticism, skepticism, and so on. So in the end science must be seen as just one of many possible convictions; as one of the ways in which thinking man deals with his existence and tries to understand the world in which he exists.

The Predicaments of Science

One of the most fascinating questions in science, one that has metaphysical overtones, is the origin and fate of the universe. There is a good deal of literature on the subject, and every now and then a statement is made about this topic, which to the ordinary man seems esoteric. A case in point is a news conference, reported by Reuter in January 1997, where two astrophysicists at the University of Michigan announced that: "The end of the universe as humans know it will occur in about 10 trillion years, but certain physical processes will continue so far into the future as astronomers can imagine" (Reuter, 1997). The first question that springs to mind immediately is that, since for untold generations such a prediction is of no consequence, why do scientists bother with it? Second, how well founded are theories on this topic? Scientists themselves, who are experts in this field, differ in their views. Some, like James Trefil, present these theories most confidently (Trefil, 1983; 1989), while others, like Stephen Hawking and Steven Weinberg, tread more carefully and with many doubts (see Chapter 5). The third point is that in making their prediction these scientists take very little risk, since there will be no scientists or any human beings to see the event and thus confirm it, if it will occur at all! And the possibilities of falsifying the theory are also limited, as we explained in Chapter 5, because of certain unverifiable assumptions and certain facts which put limits on the observability of the entire universe. So really why bother? The answer is that scientists often do science for science's sake. They seek out the frontiers of knowledge. Their intellectual curiosity leads them to questions, speculations and theories which to the layman seem imaginary and unimportant, but which from the point of view of science could lead to a better understanding of processes and events in the known and observable physical universe.

Science seems to be circumscribed not only by its inability to cope fully with the vastness and complexity of the world, but also by the fact that the latter exhibits a mixture of characteristics: order, regularity, randomness and lack of uniformity (see Chapter 2, pp. 24–30; Chapter 3, pp. 50–54; and Chapter 6, pp. 112–114). Let us remind ourselves that order and regularity are found in nature, such as in the basic chemical substances identified and ordered by the weights of their atoms, as shown by the Russian chemist Dimitry Mendeleyev as far back as 1869; or seen in the stable patterns of the social world, as for instance in economic institutions such as market-exchange systems. On the other hand, lack of regularity is also found in repeated measurements of occurrences in the natural world, and strong fluctuations occur in social, economic and political events in the social world. In the latter the sometimes sharp ups-and-downs, for instance on the stock market, may be due to both the actual changes in existing conditions and to the changing expectations and consequent speculations of economic agents.

As far as the natural world is concerned, one can suggest that irregularity and chance are basic to it, as seen in the small excess of male over female births every year, or in the infinitesimal slowing of the Earth's rotation and thus the need to add 'leap seconds' to 'universal time' (Associated Press, 1995). Then there is the fact of measurement errors which arise despite the phenomenal accuracy achieved for instance by atomic clocks, and the fantastically high resolution seen through the finest quality lenses of instruments used for observing biological structures. Errors could be due either to an inherent irregularity in nature or to the actual process of measurement. Thus, science cannot tell us exactly what aberrations there might be in the interactions and forces constituting matter and how results are affected by the application of measuring devices by human observers. Whatever we attribute it to, the occurrence of 'random errors' seems unavoidable (Gigerenzer, 1989). Admitting that such irregularities and errors occurred meant that science had to seek its basic laws in statistical probabilities. In thermodynamics the irregular and chaotic motions of billions of molecules thus led to laws showing the regularities of large numbers. Similarly, the decay or disintegration of radioactive atoms shows a statistical distribution and obeys the laws of probability when large samples are considered.

All the sciences have to live with probability data, and have to unravel the causal relationships reflected in such data. It is surprising, therefore, that some scientists of the natural world still insist on labelling their disciplines as 'the exact sciences'. They are in fact using a misnomer.

Having mentioned some of the uncertainties in science, I wish to conclude by focusing briefly on its even more fundamental predicaments. As we have seen above, science is not purely empirical and it is not culture-neutral. There are certain things science takes for granted, seen in its *a priori* assumptions. In a sense it operates as a system of beliefs, and most scientific knowledge is produced within socially institutionalized frameworks. Science is based on its own convictions, and as such it does not differ from other convictions, be they religious or ideological convictions. Its two most important premises, that of being empirically based and of arriving at its conclusions by rationality are far from being foolproof. If science takes the empiricist stance, holding that scientific knowledge is based solely on experience and facts, then it will soon be realized that this is not tenable since science cannot pull itself up by its own empiricist bootstraps; the personal judgment of the scientist and his culturally shaped presuppositions are necessary for coherently apprehending empirical reality (Polanyi, 1946). And as far as the rationalist stance is concerned, seeing the faculty of reason as its main source and test of scientific knowledge, science cannot claim that rationality can provide it with unlimited access to the understanding of reality. All thought, including rational thought, is subject to the limitations of the human intellect. And since science exists only through the human 'knower' it, too, is subject to limitations.

Quite obviously the key factor in the confrontation between science and reality *is* the human intellect. Scientists have come up with stupendous achievements during this century, yet the mystery of the awesome universe remains. Achievements which revolutionized science, such as the ability to manipulate subatomic particles, and in the biological field to carry out genetic engineering, produced moods of exalted confidence, only to be dampened by the inability of scientists to answer such basic questions as to how mass is generated or how living organisms originated (see Chapter 2, p.32 and Chapter 5, p.89). This is because the human intellect cannot grasp all that is necessary to arrive at a complete science of the entire universe. Science can only look at fragments of the universe; its conclusions are "essentially partial and temporary" (*Encyclopaedia Britannica*, 1957; *The New Encyclopaedia Britannica*, 1992). This is not to deny the heights that the human intellect can reach, seen for instance in its ability to conceive such abstract notions as infinity and eternity, or in its tendency to come up with endless questions and propositions regarding the infinite variety of combinations and causations of real events. But the idea that scientists may eventually uncover this infinite and complex reality in its totality, to achieve a kind of "Laplacean universal knowledge" (Polanyi, 1958), is a delusion. For one thing, if such could ever be achieved it would spell the end of science. But no fear: science, it seems, will forever try to fill in the gaps, to find the missing links, to achieve more knowledge and to arrive at a better understanding of reality, so that its future is guaranteed.

NOTES

[1] I have used the words 'universe', 'world' and 'reality' interchangeably in the sense of a totality, and preferred the use of one or the other depending on the context, such as when referring to the *physical universe* or the cosmos, or when distinguishing between the *physical world* and the *human world*, or when referring to *reality* as that 'whole' of things, events and human acts which we experience or which exists.

[2] I raise here certain basic questions in physics and economics in order to tell the reader that I use the term 'science' inclusively, to cover the social sciences as well as the natural sciences. It is true, however, that some claims which form part of the scientific worldview are said to apply only, or with greater force, to what are sometimes described as the 'exact sciences', such as physics. See reference to the exact sciences, above, p. 149.

BIBLIOGRAPHY

Preface

Berlin, Isaiah. *Against the Current—Essays in the History of Ideas*. Oxford: Oxford University Press, 1981, p. 80.

Newton, Roger G. *The Truth of Science—Physical Theories and Reality*. Cambridge, MA: Harvard University Press, 1997, pp. 3, 7.

Chapter One
THE DEVELOPMENT OF MODERN SCIENCE

Barkan, Elazar. *The Retreat of Scientific Racism*. Cambridge, Cambridge University Press, 1992, pp. 49n, 142, 151-161.

Bohr, Niels. *Atomic Physics and Human Knowledge*. New York: John Wiley, 1958.

Broad, C.D. *The Mind and its Place in Nature*. Paterson, N.J.: Littlefield-Adams, 1960, p. 137.

Casti, John L. *Searching for Certainty—What Scientists Can Know About the Future*. London: Abacus, 1993, pp. 33–35.

Churchland, Paul M. "A Deeper Unity." In Ronald N. Giere (ed.), *Cognitive Models of Science*. Minneapolis: University of Minnesota Press, 1992.

d'Abro, A. *The Evolution of Scientific Thought*. New York: Dover Publications, 1950.

Encyclopaedia Britannica. "Francis Bacon". Vol. 2. Chicago: 1957, pp. 882–888, 892.

(The New) Encyclopaedia Britannica, "Francis Bacon". *Macropaedia*, Vol. 2. Chicago: 1976, pp. 561–567.

Ferdon, Edwin N. "Polynesian Origins." In Andrew P. Vayda (ed.), *Peoples and Cultures of the Pacific*. Garden City, N.Y.: The Natural History Press, 1968, pp. 95–111.

Gregory, Richard L. *Mind in Science.* London: Weidenfeld and Nicolson, 1981.

Hawking, Stephen W. *A Brief History of Time.* New York: Bantam Books, 1989.

Heisenberg, Werner. *The Physicist's Conception of Nature.* New York: Hutchinson, 1958.

Heyerdahl, Thor. *The Kon-Tiki Expedition.* Harmondsworth: Penguin Books, 1963.

Kemeny, John G. *A Philosopher Looks at Science.* Princeton, N.J.: D. Van Nostrand, 1959, pp. 13, 176–178.

Kuhn, T.S. *The Structure of Scientific Revolutions.* Chicago: University of Chicago Press, 1970.

O'Hear, Anthony. *An Introduction to the Philosophy of Science.* Oxford: Clarendon Press, 1989, pp. 18, 19.

Pearson, Karl. *The Grammar of Science.* London: J.M. Dent, 1937, pp. 11, 12, 80.

Popper, Karl R. *The Logic of Scientific Discovery.* London: Hutchinson, 1959.

Prado, C.G. *The Limits of Pragmatism.* Atlantic Highlands, N.J.: Humanities Press International, 1987, pp. 99, 105.

Price, H.H. *Perception.* London: Methuen, 1950, pp. 12, 105.

Rosenthal, David M. *Materialism and the Mind-Body Problem.* Englewood Cliffs, N.J.: Prentice-Hall, 1971, p. 57.

Russell, Bertrand. *Selected Papers.* New York: The Modern Library, 1927.

Scriven, Michael. "Explanations, Predictions, and Laws." In Joseph C. Pitt (ed.), *Theories of Explanation.* New York/Oxford: Oxford University Press, 1988, pp. 57–60.

Smart, J.J.C. "Ryle in Relation to Modern Science." In Oscar P. Wood and George Pitcher (eds.), *Ryle.* London: Macmillan, 1971, pp. 283–306.

Ziman, John. *Reliable Knowledge—An Exploration of the Grounds for Belief in Science.* Cambridge: Cambridge University Press, 1978, p. 97.

Chapter Two
MAKING SENSE OF REALITY

Asimov, Isaac. *Asimov's Biographical Encyclopedia of Science and Technology.* London: George Allen and Unwin, 1966, p. 247.

Aviezer, Nathan. *In the Beginning—Biblical Creation and Science.* Hoboken, N.J.: Ktav Publishing House, 1990, pp. 67, 70.

Barnett, Lincoln. *The Universe and Dr. Einstein.* New York: Bantam, 1968, pp. 65ff.

Bohm, David. *Wholeness and the Implicate Order.* London: Routledge and Kegan Paul, 1980, p. 48.

Bohr, Niels. *Atomic Physics and Human Knowledge.* New York: John Wiley, 1958, pp. 4f., 97.

Broad, C.D. *The Mind and Its Place in Nature.* Paterson, N.J.: Littlefield Adams, 1960, pp. 81–95.

Bronowski, J. *The Ascent of Man.* Boston: Little Brown, 1973, p. 248.

Brush, Stephen G. "The Kind of Motion We Call Heat". *Physics and the Atomists,* Vol. 1, Chapter 8. Amsterdam and New York: North Holland Publishing Company, 1976, pp. 274–99.

Cairns-Smith, A.G. *Seven Clues to the Origin of Life.* Cambridge: Cambridge University Press, 1985.

d'Abro, A. *The Evolution of Scientific Thought.* New York: Dover Publications, 1950, p. 203.

Dawkins, Richard. *The Blind Watchmaker.* London: Longman, 1968; New York: Penguin, 1988, pp. 127, 148, 151.

Eisenberg, David and Donald Crothers. *Physical Chemistry.* Menlo Park, CA: The Benjamin Cummings Publishing Company, 1979, pp. 87, 114.

Encyclopaedia Britannica, Micropaedia. Chicago, 1992. Vol. 4, p. 511; Vol. 8, pp. 347, 961; Vol. 9, pp. 401–405.

Gleick, James. *Chaos—Making a New Science.* New York: Abacus, 1993, pp. 48, 122, 127, 136.

Goldstein, Martin and Inge F. Goldstein. *How We Know.* Berkeley: University of California Press, 1985, pp. 92, 106, 111.

Goerner, Sally. "Chaos, Evolution and Deep Ecology." In Robin Robertson and Allan Combs (eds.), *Chaos Theory in Psychology and the Life Sciences.* Mahwah, N.J.: Lawrence Erlbaum Associates, Publishers, 1995, pp. 18–26.

Hawking, Stephen W. *A Brief History of Time.* New York: Bantam Books, 1988, pp. 70, 85f., 108, 110, 194.

Krausz, Ernest, "Physics of Disordered Systems." *Philosophical Magazine B*, Vol. 56, No. 6 (December 1987): p. 659.

Mott-Smith, Morton. *The Concept of Energy Simply Explained.* New York: Dover Publications, 1964, pp. 155–6; 161–3, 169.

Munitz, Milton K. *Cosmic Understanding—Philosophy and Science of the Universe*, Princeton, N.J.: Princeton University Press, 1986, p. 120.

Popper, Karl R. *Objective Knowledge: An Evolutionary Approach*,. Oxford: The Clarendon Press, 1972, pp. 158ff.

Prigogine, Ilya. *The End of Certainty.* New York: Free Press, 1997.

Prigogine, Ilya and Isabelle Stengers. *Order Out of Chaos.* New York: Bantam Books, 1984, pp. xvi, 13, 117, 147–48, 154.

Rebek, Julius. "Synthetic Self-Replicating Molecules." *Scientific American*, July 1994, p. 34.

Schroeder, Gerald L. *Genesis and the Big Bang.* New York: Bantam Books, 1990, p. 25.

Singer, Charles. *A Short History of Scientific Ideas to 1900.* Oxford: Oxford University Press, 1959, pp. 352, 376, 422–425.

Stevenson, Leslie and Henry Byerly. *The Many Faces of Science.* Boulder, CO: Westview Press, 1995, p. 125.

't Hooft, Gerard, "Questioning the answers or stumbling upon good and bad theories of everything." In Jan Hilgevoord (ed.), *Physics and Our View of the World.* New York: Cambridge University Press, 1994, Chapter 1.

Trivers, Howard. *The Rhythm of Being—A Study of Temporality.* New York: Philosophical Library, 1985, p. 42.

Weinberg, Steven. *Dreams of a Final Theory.* London: Hutchinson-Radius, 1993.

Zernike, J. *Entropy.* Deventer: Kluwer, 1972, Chapter 1.

Chapter Three
SEARCHING FOR CAUSES

Aviezer, Nathan. *In the Beginning*. Hoboken, N.J.: Ktav Publishing House, 1990, p. 55.

Ayer, Alfred J. *The Foundations of Empirical Knowledge*. London: Macmillan, 1962, p. 179.

———. *Probability and Evidence*. London: Macmillan, 1972, p. 134.

Bhasker, Roy. *A Realist Theory of Science*. New York: Harvester Wheatsheaf, 1978.

Bigelow, John and Robert Pargetter. *Science and Necessity*. Cambridge/New York: Cambridge University Press, 1990, pp. 275–279, 299, 310.

Bunge, Mario. *The Myth of Simplicity—Problems of Scientific Philosophy*. Englewood Cliffs, N.J.: Prentice Hall, 1963, pp. 182, 195, 203.

Cantor, Charles R. and Paul R. Schimmel. *Biophysical Chemistry—Part I: The Conformation of Biological Macromolecules*. San Francisco: W.H. Freeman, 1980, p. 79.

Casti, John L. *Searching for Certainty*. London: Abacus/Little Brown, 1993, Chapter 1.

Cline, David B. "Low-Energy Ways to Observe High-Energy Phenomena." *Scientific American*, Vol. 271, No. 2 (September 1994), p. 26.

(A) Dictionary of Philosophy. Antony Flew (Consultant ed.). London: Pan Books/The Macmillan Press, 1979, p. 78.

Dunbar, Robin. *The Trouble with Science*. London: Faber and Faber, 1995.

Dupre, John. "A Probabilistic Causality." *Philosophy of Science*, Vol. 57, No. 4 (1990), p. 690.

Funkenstein, Amos. "The Persecution of Absolutes: On the Kantian and Neo-Kantian Theories of Science." In Edna Ullman-Margalit (ed.), *The Kaleidoscope of Science*, Dordrecht, Holland: D. Reidel, 1986, pp. 40, 51.

Gould, S.J. *Conceptual Issues in Evolutionary Biology*. E. Sober (ed.). Cambridge, Massachusetts: MIT Press, 1984, pp. 256, 260, quoted in Nathan Aviezer, *In the Beginning*. Hoboken, N.J.: Ktav Publishing House, 1990, pp. 54–59; see also *New Scientist*, 8th April, 1995.

Hanson, Norwood Russell. *Patterns of Discovery.* Cambridge: Cambridge University Press, 1961, p. 50.

Hawking, Stephen W. *A Brief History of Time.* New York: Bantam Books, 1988, pp. 34, 75.

Heise, David R. *Causal Analysis.* New York: John Wiley, 1975, pp. 27, 30, 31, 42, 46.

Horgan, John. "Revisiting Old Battlefields." *Scientific American*, Vol. 270, No. 4 (April, 1994), pp. 19–21.

Hospers, John. *An Introduction to Philosophical Analysis.* London: Routledge, 1956, pp. 223–249.

Iannotta, Ben. "Music of the Spheres." *New Scientist.* London, 31 August 1996, p. 28.

Kockelmans, Joseph J. (ed.). *Philosophy of Science—The Historical Background.* New York/London: The Free Press/Collier Macmillan, 1968, pp. 67, 68, 70.

Lieberson, Stanley. "Einstein, Renoir, and Greeley, "Some Thoughts about Evidence in Sociology." *American Sociological Review*, Vol. 57, No. 1 (1992), p. 5.

McGraw-Hill Yearbook of Science and Technology 1995. Sybil P. Parker (Editor-in-Chief). New York: McGraw-Hill, 1994.

Milton, Richard. *The Facts of Life: Shattering the Myths of Darwinism.* London: Corgi Books, 1992.

O'Hear, Anthony. *An Introduction to the Philosophy of Science.* Oxford: Clarendon Press, 1989, p. 91.

Palter, Robert M. *Whitehead's Philosophy of Science.* Chicago/London: The University of Chicago Press, 1960, p. 160f.

Pearson, Karl. *The Grammar of Science.* London: J.M. Dent, 1937, pp. 107–109.

(The) Penguin Dictionary of Science. E.B. Uvarov and Alan Isaacs (eds.), Seventh Edition. London: Penguin Books, 1993, pp. 153, 191, 378.

Popper, Karl R. *The Logic of Scientific Discovery.* London: Hutchinson, 1975.

Rescher, Nicholas. *Scientific Explanation.* New York: The Free Press, 1970, p. 125.

Russell, Bertrand. *Human Knowledge, Its Scope and Limits.* New York: Simon and Schuster, 1948, p. 59.

Salmon, Wesley C. "Statistical Explanation and Causality." In Joseph C. Pitt (ed.), *Theories of Explanation*, New York/London: Oxford University Press, 1988, pp. 82, 96.

Simon, Herbert A. *Models of Discovery*. Dordrecht, Holland/Boston, USA: D. Reidel, 1997, pp. 55f., 95.

Weisskopf, Victor F. "Frontiers and Limits of Science." The Joseph Wunsch Lecture, Technion-Israel Institute of Technology, Haifa, March 23, 1981, p. 6.

Whitehead, Alfred North. *The Concept of Nature*. Cambridge: Cambridge University Press, 1920, 1964 edition, pp. 46–48.

Chapter Four
CHANCE, IMAGINATION, DECEPTION

Anthropology Today. Del Mar, CA: CRM Books, 1971, pp. 10, 24.

Ashall, Frank. *Remarkable Discoveries*. Cambridge: Cambridge University Press, 1994, p. 33.

Aviezer, Nathan. *In the Beginning*. Hoboken, N.J.: Ktav Publishing House, 1990, p. 54.

———. "Misreading the Fossils: The Dark Side of Evolutionary Biology." *BDD Journal of Torah and Scholarship*, Bar-Ilan University Press, No. 2 (Winter 1996): 23, 24, 26.

Bernal, J.D. *Science in History*, Vol. 2. Harmondsworth, Middlesex: Penguin Books, 3rd edition, 1965, pp. 610, 734f.

Bernstein, Jeremy. *Experiencing Science*. New York: Dutton, 1978, pp. 133–142.

Bronowski, J. *The Ascent of Man*. Boston/Toronto: Little Brown, 1973, p. 38.

Chalmers, A.F. *What is this thing called science?* St. Lucia, Queensland: University of Queensland Press, 1976, pp. 104–105.

Conant, James B. *On Understanding Science*. New York: Mentor/Yale University Press, 1947, pp. 27, 102.

Crowther, J.G. *British Scientists of the Nineteenth Century*. London: Routledge and Kegan Paul, 1935, pp. 101–105.

Domb, Cyril, "Torah and Science—Are they Converging?" *BDD Journal of Torah and Scholarship*, Bar-Ilan University Press, No. 1 (Summer 1995): 38.

The New Encyclopaedia Britannica, Micropaedia. Chicago, Vol. 9, 1992, p. 445, and Vol. 1, 1992, p. 526.

Freeman, D. *Margaret Mead and Samoa: The Making and Unmaking of an Anthropological Myth.* Cambridge, MA: Harvard University Press, 1983.

Gillispie, Charles Coulston. *The Edge of Objectivity—An Essay in the History of Scientific Ideas*. Princeton, N.J.: Princeton University Press, 1960, pp. 204-213, 491.

Gould, S.J. *The Panda's Thumb.* London: Penguin, 1980, p. 97.

Harre, Rom. *Great Scientific Experiments.* Oxford: Oxford University Press, 1983, pp. 96–103, 108–112.

Huizenga, John. *Cold Fusion—The Scientific Fiasco of the Century.* Rochester, N.Y.: University of Rochester Press, 1992, pp. x-xi, 8, 22–23, 25, 34, 79, 108, 164–198, 203–236.

Jacob, Francois. "Review of the Possible and the Actual." In Perutz, M. (ed.). *Is Science Necessary?* London: Barry and Jenkins, 1989, p. 210.

Kamin, L. *The Science and Politics of IQ.* Potomac, MD: Erlbaum, 1974.

Kockelmans, Joseph J., (ed.). *Philosophy of Science.* New York/London: The Free Press/Collier MacMillan, 1968, pp. 12–15.

Kohn, Alexander. *False Prophets.* Oxford: Basil Blackwell, 1986, pp. 18–20, 52–57, 117–120, 141–143.

———. *Fortune or Failure.* Oxford: Basil Blackwell, 1989, pp. 1, 119, 168–170.

Lipetz, Ben-Ami. *The Measurement of Efficiency of Scientific Research.* Carlisle, MA: Intermedia, 1965, pp. 78–79.

McKusick, Victor A. "The Human Genome Organization: History, Purposes and Membership." *Genomics*, Vol. 5 (1989): pp. 385–387.

Mead, M. *Coming of Age in Samoa.* New York: Blue Ribbon Books, 1933.

Medawar, Peter, *The Limits of Science*, Oxford University Press, Oxford, 1984, pp. 83–87.

Parker, Sybil P. (ed.). *McGraw-Hill Dictionary of Scientific and Technical Terms.* New York: McGraw-Hill Book Company, 3rd edition, 1984.

Pearson, Karl. *The Grammar of Science.* London: Dent, 1937 edition, p. 219.

Perutz, Max F. *Is Science Necessary?* London: Barry and Jenkins, 1989, pp. 197, 210.

Ravetz, J.R. *Scientific Knowledge and its Social Problems*, Oxford: Oxford University Press, 1971, p. 81.

Reuter, Linus Pauling obituary, 21.8.1994.

Rose, Paul Lawrence. *Heisenberg and the Nazi Atomic Bomb Project.* Berkley, Los Angeles, London: University of California Press, 1998.

Schroeder, Gerald L. *Genesis and the Big Bang,* New York: Bantam Books, 1990, pp. 25, 37f., 177f.

Science Year 1991. The World Book Annual Science Supplement. Chicago: World Book, 1991, pp. 325–26.

Taton, R. *Reason and Chance in Scientific Discovery.* London: The Scientific Book Guild-Beaverbrook Newspapers/Hutchinson, 1957, pp. 79, 85–86.

Thomson, George. *The Inspiration of Science.*, London: Oxford University Press, 1961, p.1.

Toulmin, Stephen and June Goodfield. *The Fabric of the Heavens.* New York: Harper and Row, 1961, p. 257.

Trinkaus, E. *The Neanderthals.* London: Jonathan Cape, 1993.

Wade, N. "IQ and heredity: suspicion of fraud beclouds classic experiment." *Science*, 194 (1976): 916.

Chapter Five
THE COSMOS—FROM EVIDENCE TO SPECULATION

Aviezer, Nathan. *In the Beginning.* Hoboken, N.J.: Ktav Publishing House, 1990, pp. 9, 13.

Curran, Noel. *The Logical Universe—The Real Universe.* Aldershot: Avebury, 1994.

Davies, Paul. *The Mind of God.* New York: Simon and Schuster, 1992, pp. 196ff., 199, 200, 217.

Davies, P.C.W. *Space and Time in the Modern Universe.* Cambridge: Cambridge University Press, 1977, pp. 164–66.

Davies, Paul and John Gribbin. *The Matter Myth.* Harmondsworth, Middlesex: Penguin, 1992.

de Vaucouleurs, Gerard. *Discovery of the Universe*. New York: The Macmillan Company, 1957, p. 289.

Domb, Cyril, "Torah and Science—Are they Converging?" *BDD Journal of Torah and Scholarship* Bar-Ilan University Press, No. 1 (Summer 1995): 36.

Drees, Willem B. "Problems in debates about physics and religion." In Jan Hilgevoord (ed.), *Physics and Our View of the World*. Cambridge: Cambridge University Press, 1994, p. 204.

The Economist. Vol. 338, No. 7949. London, January 20-26, 1996, p. 90.

Gribbin, John. "Hot Dark Matter Loses its Mystery." *New Scientist*, London, 1995.

Guth, Allan H., "Speculations on the Origin of Matter, Energy and Entropy in the Universe." In A.H. Guth, K. Huang and R.L. Jaffe (eds.), *Asymptotic Realms of Physics*. Cambridge, MA: MIT Press, 1983.

Hawking, Stephen W. *A Brief History of Time*. New York: Bantam Books, pp.50, 51.

Hey, Tony and Patrick Walters. *The Quantum Universe*. Cambridge: Cambridge University Press, 1987.

Kaveh, Moshe. *Hatzofeh*, Tel-Aviv, 8.7.1994 (Hebrew).

Midrashim—Bereshit Rabba. Chapter 3, Section 7; Chapter 9, Section 2.

Munitz, Milton K. *Space, Time and Creation—Philosophical Aspects of Scientific Cosmology*. New York: Dover Pulblications, Inc. (2nd edition), 1981.

———. *Cosmic Understanding*. Princeton, N.J.: Princeton University Press, 1986, pp. 6, 50, 116–17, 125, 150, 152–154.

Olby, R.C., et al. (eds.). *Companion to the History of Modern Science*. London/New York: Routledge, 1990.

Parker, Sybil P. (ed.). *McGraw-Hill Yearbook of Science and Technology* 1995. New York: McGraw-Hill Book Company, 1994, pp. 186–87.

Perutz, Max F. *Is Science Necessary?*. London: Barry and Jenkins, 1989, p. 17.

Sakurai, J.J. *Modern Quantum Mechanics*. Menlo Park, CA: The Benjamin/Cummings Publishing Co., Inc., 1985, pp. 226–27.

Schroeder, Gerald L. *Genesis and the Big Bang*, New York: Bantam Books, 1990, pp. 24, 42, 71–73, 78–79, 83.

Shapere, Dudley. *Reason and the Search for Knowledge*. Dordrecht, Holland: D. Reidel, 1984, p. 176, Chapters 9, 11.

't Hooft, Gerard. "Questioning the answers or stumbling upon good and bad theories of everything." In Jan Hilgevoord (ed.), *Physics and Our View of the World*. Cambridge: Cambridge University Press, 1994, p. 35.

Trivers, Howard. *The Rhythm of Being—A Study of Temporality*. New York: Philosophical Library, 1985, pp. 33, 39.

van Fraassen, Bas C. *Quantum Mechanics: An Empiricist View*. Oxford: Clarendon Press, 1991, pp. 338-39.

Weinberg, Steven. *The First Three Minutes*. Glasgow: André Deutsch, 1977; Fontana, 1978; Flamingo Edition, 1983, pp. 30–33, 63, 76, 118–19, 143–45, 148, 160.

———. *Dreams of a Final Theory*. London: Hutchinson Radius; New York: Random House Vintage Books, 1993.

Zohar. Appendix, pp. 252/12, 253/1.

Chapter Six
IS 'SOCIAL SCIENCE' REALLY SCIENCE?

Allais, Maurice. "Contributions to Economics." In David L. Sills (ed.), *International Encyclopedia of the Social Sciences*, New York: The Macmillan Company and The Free Press, Vol. 2, 1968, pp. 403–404.

Bauman, Z. *Hermeneutics and Social Science*. New York: Columbia University Press, 1978, p. 12.

Berent, Enc. "Units of Behavior." *Philosophy of Science*, Vol. 62, No 4 (Dec. 1995): 523–542.

Blalock, H.M. "The Measurement Problem." In H.M. Blalock and A.B. Blalock (eds.), *Methodology in Social Research*, New York: McGraw-Hill, 1968.

Blaug, Mark. In *International Encyclopedia of the Social Sciences*, David L. Sills (ed.). New York: The Macmillan Company and The Free Press, Vol. 13, 1968, pp. 507–511.

Boudon, Raymond. *The Logic of Social Action*. London: Routledge, 1981.

———. "Individual action and social change: A no-theory of social change." *The British Journal of Sociology*, Vol. 34, 1983, pp. 1–18.

Bridgman, P.W. *The Logic of Modern Physics.* New York: Macmillan, 1928.

Casti, John, L. *Searching for Certainty.* London: Abacus, 1993, p.231.

Coleman, James S. *Foundations of Social Theory.* Cambridge, MA: The Belknap Press of Harvard University Press, 1990, pp. 2, 37ff.

Coleman, James S., and Thomas J. Fararo (eds.). *Rational Choice Theory.* London: Sage Publications, 1992.

Corry, Bernard. In David L. Sills (ed.), *International Encyclopedia of the Social Sciences.* New York: The Macmillan Company and The Free Press, Vol. 10, 1968, pp. 25–32.

Dennett, Daniel C. "True believers: the intentional strategy and why it works." In A.F. Heath (ed.), *Scientific Explanation.* Oxford: Clarendon Press, 1981, pp. 64, 65.

Durkheim, Emile. *Rules of Sociological Method,* translated by S.A. Solvay and J.H. Mueller. Chicago: University of Chicago Press, 1938.

———. "The External Reality of Society." In Edgar F. Borgatta and Henry J. Meyer (eds.), *Sociological Theory: Present Day Sociology from the Past.* New York: Knopf, 1956.

(The) Economist, London, 14th October, 1995, p. 118.

(The New) Encyclopaedia Britannica, Macropaedia, Vol. 12, Chicago, 1974, p. 224.

Fodor, Jerry A. *The Language of Thought.*, Hassocks, Sussex: The Harvester Press, 1976, p.16.

Freedman, R., et. al. *Principles of Sociology.* New York: Henry Holt, 1956.

Frey, Bruno S. *Economics as a Science of Human Behavior.* Boston/Dodrecht/London: Kluwer Academic Publishers, 1992, p. 171.

Galbraith, J.K. "The Good Society." *The Guardian 2,* London/ Manchester, 26th January, 1994.

Giddens, Anthony. *Profiles and Critique in Social Theory.* Berkeley and Los Angeles: University of California Press, 1982, p.8.

———. *The Constitution of Society: Outline of the Theory of Structuration.* Berkeley and Los Angeles: University of California Press, 1984.

Gould, S.J. *The Rational Society.* Auguste Comte Memorial Trust Lecture 8, University of London. London: The Athlone Press, 1971.

Harbury, C.D. *An Introduction to Economic Behavior*. London: Fontana/Collins, 1971, p. 140.

Hausman, Daniel M. (ed.). *The Philosophy of Economics* (Second Edition). Cambridge: Cambridge University Press, 1994, pp.29, 33, 34.

Hillman, Arye L., "The transition from socialism: An overview from a political-economy perspective", *European Journal of Political Economy*, Vol. 10, 1994, pp. 191–225.

———. "Western
Economic Theory and the Transition: The Public Choice Perspective." *Economics and Mathematical Methods* (Journal of the Central Economic and Mathematical Institute of the Russian Academy of Sciences), 1996, pp. 4–12.

Homans, George C. "The present state of sociological theory." *The Sociological Quarterly*, Vol. 23, 1982, pp. 285–299.

Kirzner, Israel M. *The Economic Point of View*. Kansas City: Sheed and Ward, 1960, pp. 23, 33, 56–57, 143.

Krausz, Ernest. *Sociological Research—A Philosophy of Science Perspective*. Assen/Maastricht: Van Gorcum, 1986, Chapters 2 and 4.

———. "Leaving Man in Society." *Methodology and Science*, Vol. 25, No. 3, 1992, p. 149.

Krausz, E., and S.H. Miller. *Social Research Design*. London: Longman, 1974, pp. 8. 16.

Lundberg, G.A. *Foundations of Sociology*. New York: Macmillan, 1939.

Luria, A.R. *The Working Brain—An Introduction to Neuropsychology*, translated by Basil Haigh. London: Allen Lane/The Penguin Press, 1973, pp. 16, 42, 340.

MacLennan, Bruce. "The Elements of Consciousness and their Neuro-dynamical Correlates." *Journal of Consciousness Studies*, Vol. 3, Nos. 5–6, 1996, pp. 409–424.

Maital, Shlomo. *Minds, Markets, and Money*. New York: Basic Books, 1982, p. 21.

Marschak, Jacob. "On Econometric Tools." Hausman, Daniel M. (ed.). *The Philosophy of Economics* (Second Edition). Cambridge: Cambridge University Press, 1994, Chapter 15.

McKenzie, Richard B., and Gordon Tullock. *Modern Political Economy*. New York: McGraw-Hill, 1978, Chapter 2.

Mennell, Stephen. *Sociological Theory*. New York: Praeger, 1974, pp. 8–10, 13–15.

Miller, Roger LeRoy and Robert W. Pulsinelli. *Modern Money and Banking*. New York: McGraw-Hill, 1985, pp. 339–340.

Mitchell, Duncan. *A Dictionary of Sociology*. London: Routledge and Kegan Paul, 1968, p.218.

Morgenstern, Oskar. "Game Theory." In David L. Sills (ed.), *International Encyclopedia of the Social Sciences*, New York: The Macmillan Company and The Free Press, Vol. 6, 1968, pp. 62–68.

Mueller, Dennis C., *Public Choice*, Cambridge, Cambridge University Press, 1979, pp. 1,3.

———. "The corporation and the economist." In Hausman, Daniel M. (ed.). *The Philosophy of Economics* (Second Edition). Cambridge: Cambridge University Press, 1994, p. 308.

Olson, Mancur. "The Political Economy of Comparative Growth Rates." In Dennis C. Mueller (ed.), *The Political Economy of Growth* New Haven and London: Yale University Press, 1983.

Osborne, Martin J. and Ariel Rubinstein, *A Course in Game Theory*. Cambridge, MA: The MIT Press, 1994, p.4.

Penrose, Roger. *Shadows of the Mind—A Search for the Missing Science of Consciousness*. Oxford: Oxford University Press, 1994, pp. 39, 367, 371, 411.

Pepperell, Robert. *The Post-Human Condition*. Oxford: Intellect, 1995, pp. 4–12.

Phillips, D.C. *The Social Scientist's Bestiary*. Oxford: Pergamon Press, 1992, pp. 2–5, 96–97, 98, 104.

Polanyi, Michael. "On the Modern Mind." In Fred Schwartz (ed.), *Scientific Thought and Social Reality: Essays by Michael Polanyi*, Psychological Issues, Vol. VIII, No. 4, Monograph 32. New York: International Universities Press, 1974, pp. 138–39.

Polkinghorne, John. *Beyond Science*, Cambridge, Cambridge University Press – Canto Edition, 1998, pp. 59-69.

Price-Williams, D.R., "Cognition", in J. Gould and W.L. Kolb, eds., *A Dictionary of the Social Sciences*, Tavistock, London, 1964, p.89.

Rapaport, Steven. "Economic Models and Historical Explanation." *Philosophy of the Social Sciences*, Vol. 25, No. 4, December 1995, pp. 421–441.

Revonsuo, Antii and Matti Kamppinen (eds.). *Consciousness in Philosophy and Cognitive Neuroscience*. Hillsdale, N.J.: Lawrence Erlbaum Associates, 1994, pp. 1–5, 8–15, 20, 26, 230.

Robbins, L. *An Essay on the Nature and Significance of Economic Science* (Second Edition). London: Macmillan, 1935, p.16.

Rose, A.M. and C.B. Rose. *Sociology: The Study of Human Relations*. New York: Knopf, 1965.

Rosenberg, Alexander. "If economics isn't science, what is it?" In Hausman, Daniel M. (ed.), *The Philosophy of Economics* (Second Edition). Cambridge: Cambridge University Press, 1994, pp. 376, 382.

Ross, Dorothy. *The Origins of American Social Science*. Cambridge: Cambridge University Press, 1991, p. 367.

Rotwein, Eugene. In David L. Sills (ed.), *International Encyclopedia of the Social Sciences*. New York: The Macmillan Company and The Free Press, Vol. 6, 1968, pp. 546–550.

Scrapanti, E. and S. Zamagni. *An Outline of the History of Economic Thought*. Oxford: Clarendon Press, 1995, p. 234.

Seamon, John G. and Douglas T. Kenrick. *Psychology*. Englewood Cliffs, N.J.: Prentice-Hall, 1992, pp. 9, 10.

Seger, Imogen. *Introduction to Sociology*. London: Rupert Hart-Davis, 1972, pp. 7, 13.

Schumpeter, Joseph. "Science and Ideology." In Hausman, Daniel M. (ed.). *The Philosophy of Economics* (Second Edition). Cambridge: Cambridge University Press, 1994, p. 235.

Sichel, Werner (ed.). *The State of Economic Science*. Kalamazoo, MI: W.E. Upjohn Institute, 1989, pp. 25, 32, 20–21.

Skinner, B.F. *Science and Human Behavior*. New York: Macmillan, 1953.

———. *Beyond Freedom and Dignity*. New York: Knopf, 1971, p.8.

Stevens, S.S. "Psychology and the science of science." *Psychological Bulletin*, Vol. 36, 1939, pp. 221–63.

Stonier, Alfred W. and Douglas C. Hague. *A Textbook of Economic Theory*. London: Longman, Green and Co., 1953, Chapters 1–3.

Tulea, Gitta and Ernest Krausz. "Changing Approaches in Postmodern Sociological Thought." *International Journal of Comparative Sociology*, Vol. 34, Nos. 3–4, December 1993, p. 215.

Turner, Stephen P. "On the relevance of statistical relevance theory." *Theory and Decision*, Vol. 14, 1982, pp. 195–205.

Tversky, Amos and Daniel Kahneman. "Rational Choice and the Framing of Decisions." In David E. Bell, Howard Raiffa and Amos Tversky (eds.), *Decision Making*. Cambridge: Cambridge University Press, 1988, p.172.

Vernon, Philip A. *The Neuropsychology of Individual Differences*. San Diego, CA: Academic Press, 1994, p. xiii.

Wallerstein, Immanuel et. al. *Open the Social Sciences*, Report of the Gulbenkian Commission on the Restructuring of the Social Sciences. Stanford, CA: Stanford University Press, 1996.

Wallich, Paul. "Not worth the paper it's printed on." *Scientific American*, September 1994, p.89.

Watson, J.B. *Behaviorism*. New York: The People's Institute Publishing Co. 1925.

Wicksteed, Philip H. *The Common Sense of Political Economy*. London: George Routledge and Sons, 1933.

Wisdom, J.O. *Philosophy of the Social Sciences: A Metascientific Introduction*. Aldershot: Avebury, 1987, pp. 5, 6.

Chapter Seven
THE SCIENTIFIC WORLDVIEW

Associated Press (Washington). "'Leap second' will delay 1996", 17.12.1995.

Bellah, Robert N. *Beyond Belief*. Berkeley: University of California Press, 1991, pp. 243–246.

Benton, Ted. *Philosophical Foundations of Three Sociologies*. London: Routledge and Kegan Paul, 1977, pp. 65f., 71.

Cherniak, Christopher. *Minimal Rationality*. Cambridge, MA: The MIT Press, 1986, pp. 127–129.

Encyclopaedia Britannica. "Science", Vol. 20, Chicago, 1957, p.123.

(The New) Encyclopaedia Britannica, Macropaedia. "Science", Vol. 27, Chicago, 1992, p.41.

Feyerabend, Paul. *Against Method.* London: Verso, 1975, pp. 295–298.

———. *Three Dialogues on Knowledge*. Oxford: Basil Blackwell, 1991, p.164.

Gigerenzer, Gerd et. al..*The Empire of Chance.* Cambridge: Cambridge University Press, 1989, pp. 45, 79–80, 180–181.

Giddens, Anthony (ed.) *Emile Durkheim: Selected Writings.* Cambridge/London: Cambridge University Press, 1972, p. 248f.

Habermas, Jürgen. *Knowledge and Human Interests* (2nd Edition). London: Heinemann, 1978, p.4.

Hanson, N.R. *Perception and Discovery.* San Francisco: Freeman, Cooper, 1969, p.408.

Horgan, John. *The End of Science.* New York: Edison Wesley, 1996.

Knight, David. *The Age of Science—The Scientific World-View in the Nineteenth Century*. Oxford: Basil Blackwell, 1986.

Lewis, Clarence Irving. "Empiricism." In Douglas G. Arner (ed.), *Perception, Reason, and Knowledge*, Glenview, IL: Scott, Foreman and Company, 1972, p.125.

Merton, Robert K. *The Sociology of Science.* Chicago: The University of Chicago Press, 1973, pp. 264, 267–278.

Mulkay, Michael. *Science and the Sociology of Knowledge.* London: George Allen and Unwin, 1979, pp. 3-5, 6-9, 19, 20, 28.

Musgrave, Alan. *Common Sense, Science and Scepticism.* Cambridge: Cambridge University Press, 1993.

Narkhnikian, George (ed.). *Bertrand Russell's Philosophy.* New York: Harper and Row, 1974, pp. 142, 207, 219.

Planck, Max. *Scientific Autobiography and Other Papers.* New York: Philosophical Library, 1949, p.82.

Polanyi, Michael. *Science, Faith and Society.* Chicago: Phoenix Books, The University of Chicago Press, 1946, p.38.

———. *Personal Knowledge.* London: Routledge and Kegan Paul, 1958, pp. vii, viii, 140.

Polanyi, Michael and Harry Prosch. *Meaning*. Chicago: The University of Chicago Press, 1975, pp. 28–31, 56.

Popper, Karl R. *The Logic of Scientific Discovery* London: Hutchinson1959, p.111.

Redner, Harry. *The Ends of Science*. Boulder and London: Westview Press, 1987, p.282.

Reuter (Toronto), "Universe will end in 10 trillion years", 16.1.1997.

Russell, Bertrand. *The Scientific Outlook*. New York: W.W. Norton, 1931.

———. *The Impact of Science on Society*. London: Unwin, 1976, pp. 25, 102.

Ryle, Gilbert. *The Concept of Mind*. London: Hutchinson, 1966.

Sainsbury, R.M. *Russell*. London: Routledge and Kegan Paul, 1979, p.193f.

Stevenson, Leslie and Henry Byerly. *The Many Faces of Science*. Boulder, CO.: Westview Press, 1995.

Sklair, Leslie. *Organized Knowledge—A Sociological View of Science and Technology*. London: Hart-Davis/McGibbon, 1973.

Trefil, James S. *The Moment of Creation*. New York: Charles Scribner's Sons, 1983.

———. *Reading the Mind of God—In Search of the Principle of Universality*. New York: Charles Scribner's Sons, 1989, pp. 204–212.

Uvarov, E.B., and Alan Isaacs. *The Penguin Dictionary of Science*. Harmondsworth, Middlesex: Penguin Books, 1993, p.361.

Woolgar, Steve. *Science: The Very Idea*. Chichester/London: Ellis Horwood/Travistock Publications, 1988, pp. 30–33, 100f.

Ziman, J.M. *Public Knowledge* Cambridge: Cambridge University Press, 1968.

———. *Reliable Knowledge*. Cambridge: Cambridge University Press, 1978, pp. 9, 13–16, 25, 99–100, 110, 150f, 168f.

———. *An Introduction to Science Studies*. Cambridge: Cambridge University Press, 1984, pp. 35–40, 53, 84–86.

SUBJECT INDEX

AUTHOR INDEX

WPI Studies

WPI Studies is sponsored by Worcester Polytechnic Institute, the nation's third oldest independent technological university. WPI Studies aims to publish monographs, edited collections of essays, and research tools and texts of interest to scholarly audiences. WPI Studies accepts manuscripts in all languages, and is especially interested in reviewing potential publications on interdisciplinary topics relating science, technology, and culture. WPI Studies is edited by a board of WPI faculty from many disciplines. The board is chaired by Lance Schachterle, Assistant Provost and Professor of English, to whom potential authors should direct their inquiries (WPI, Worcester, MA 01609).

To order other books in this series, please contact our Customer Service Department at:

800-770-LANG (within the U.S.)
212-647-7706 (outside the U.S.)
212-647-7707 FAX

or browse online by series at:

www.peterlang.com

Kolibri 360 GmbH
Ferdinand-Jühlke-Straße 7
99095 Erfurt, Deutschland
produktsicherheit@kolibri360.de

Druck:
CPI Druckdienstleistungen GmbH
im Auftrag der
Kolibri 360 GmbH
Ferdinand-Jühlke-Str. 7
99095 Erfurt